AF595341

Enzyme Inhibitor from Marine Organisms

Enzyme Inhibitor from Marine Organisms

Editor

Dirk Tischler

MDPI • Basel • Beijing • Wuhan • Barcelona • Belgrade • Manchester • Tokyo • Cluj • Tianjin

Editor
Dirk Tischler
Ruhr-University Bochum
Germany

Editorial Office
MDPI
St. Alban-Anlage 66
4052 Basel, Switzerland

This is a reprint of articles from the Special Issue published online in the open access journal *Marine Drugs* (ISSN 1660-3397) (available at: https://www.mdpi.com/journal/marinedrugs/special_issues/Enzyme_Inhibitor).

For citation purposes, cite each article independently as indicated on the article page online and as indicated below:

LastName, A.A.; LastName, B.B.; LastName, C.C. Article Title. *Journal Name* **Year**, *Volume Number*, Page Range.

ISBN 978-3-03943-783-2 (Hbk)
ISBN 978-3-03943-784-9 (PDF)

Contents

About the Editor . . . . **vii**

Dirk Tischler
A Perspective on Enzyme Inhibitors from Marine Organisms
Reprinted from: *Mar. Drugs* **2020**, *18*, 431, doi:10.3390/md18090431 . . . . **1**

Siqi Sun, Xiaoting Xu, Xue Sun, Xiaoqian Zhang, Xinping Chen and Nianjun Xu
Preparation and Identification of ACE Inhibitory Peptides from the Marine Macroalga *Ulva intestinalis*
Reprinted from: *Mar. Drugs* **2019**, *17*, 179, doi:10.3390/md17030179 . . . . **5**

Antonio Jesús Vizcaíno, Alba Galafat, María Isabel Sáez, Tomás Francisco Martínez and Francisco Javier Alarcón
Partial Characterization of Protease Inhibitors of *Ulva ohnoi* and Their Effect on Digestive Proteases of Marine Fish
Reprinted from: *Mar. Drugs* **2020**, *18*, 319, doi:10.3390/md18060319 . . . . **23**

Pei Qiu, Zhaoming Liu, Yan Chen, Runlin Cai, Guangying Chen and Zhigang She
Secondary Metabolites with α-Glucosidase Inhibitory Activity from the Mangrove Fungus *Mycosphaerella* sp. SYSU-DZG01
Reprinted from: *Mar. Drugs* **2019**, *17*, 483, doi:10.3390/md17080483 . . . . **39**

Jinhyuk Lee and Mira Jun
Dual BACE1 and Cholinesterase Inhibitory Effects of Phlorotannins from *Ecklonia cava*—An In Vitro and in Silico Study
Reprinted from: *Mar. Drugs* **2019**, *17*, 91, doi:10.3390/md17020091 . . . . **51**

Irina Bakunina, Galina Likhatskaya, Lubov Slepchenko, Larissa Balabanova, Liudmila Tekutyeva, Oksana Son, Larisa Shubina and Tatyana Makarieva
Effect of Pentacyclic Guanidine Alkaloids from the Sponge *Monanchora pulchra* on Activity of α-Glycosidases from Marine Bacteria
Reprinted from: *Mar. Drugs* **2019**, *17*, 22, doi:10.3390/md17010022 . . . . **67**

Te Li, Ning Wang, Ting Zhang, Bin Zhang, Thavarool P. Sajeevan, Valsamma Joseph, Lorene Armstrong, Shan He, Xiaojun Yan and C. Benjamin Naman
A Systematic Review of Recently Reported Marine Derived Natural Product Kinase Inhibitors
Reprinted from: *Mar. Drugs* **2019**, *17*, 493, doi:10.3390/md17090493 . . . . **83**

About the Editor

Dirk Tischler (Prof. Dr.) has run the Microbial Biotechnology Group at Ruhr-Universität Bochum, Germany, since 2018. He studied applied natural science at the TU Bergakademie Freiberg, Germany, and obtained a diploma in 2007 and Dr. rer. nat. in 2012. From 2012 until now, he has been a group leader with a main interest in identifying novel enzymes and describing their properties for general and applied purposes.

Editorial

A Perspective on Enzyme Inhibitors from Marine Organisms

Dirk Tischler

Microbial Biotechnology, Faculty of Biology and Biotechnology, Ruhr-Universität Bochum, 44780 Bochum, Germany; dirk.tischler@rub.de; Tel.: +49-234-32-22656

Received: 6 August 2020; Accepted: 14 August 2020; Published: 19 August 2020

Abstract: Marine habitats are promising sources for the identification of novel organisms as well as natural products. Still, we lack detailed knowledge on most of the marine biosphere. In the last decade, a number of reports described the potential of identifying novel bioactive compounds or secondary metabolites from marine environments. This is, and will be, a promising source for candidate compounds in pharma research and chemical biology. In recent years, a number of novel techniques were introduced into the field, and it has become easier to actually prospect for natural products, such as enzyme inhibitors. These novel compounds then need to be characterized and evaluated in comparison to well-known representatives. A number of current research projects target the exploitation of marine organisms and thus the corresponding diversity of metabolites. These are often encountered as potential drugs or biological active compounds. Among these, the class of enzyme inhibitors is an important group of compounds. There is room for new discoveries, and some more recent discoveries are highlighted herein.

Keywords: secondary metabolites; functional annotation; structure–function relation; natural products; bioactives; enzyme inhibition; inactivation; marine bacteria; marine fungi; marine sponges

---

1. Marine Habitats as Sources of Natural Products

More than 70% of the planet Earth's surface is covered by water, and little is known about the biosphere within these habitats, often designated as marine environment. With respect to its enormous dimension, considering its huge surface area and depths to more than 10,000 m, it is obvious that the diversity of living organisms must be high in marine habitats. Actually, it is assumed that more than 80% of all organisms occur in such marine ecosystems [1]. This provides avenues to new creatures comprising a rich diverse metabolisms and thus novel metabolites can be explored or uncovered.

In the last few years, a number of original articles and reviews have reported on the potential of identifying novel bioactive compounds or secondary metabolites from marine environments, including enzyme inhibitors, which is obviously linked to the tremendous diversity of marine life. In particular, marine bacteria or fungi, cyanobacteria, corals, sponges, algae, and worms, among others, are known to produce a variety of bioactive compounds (including saccharides, polysaccharides, peptides, polyketides, polyphenolic compounds, sterol-like products, alkaloids, quinones, and quinolines, just to name a few compound classes). Often these are produced and excreted as secondary metabolites, as, for example, well-described antimicrobial compounds against Gram-positive bacteria isolated from several marine *Streptomyces* or those of sediments from such ecosystems. These antimicrobial active agents are especially interesting in the context of the still ongoing development of multi-resistant strains, which is critical for our current lifestyle. Anthracimycin [2–4] and aplasmomycin [5] are just two representatives to be mentioned that act against Gram-positive bacteria [1]. DNA and RNA biosynthesis are inhibited due to anthracimycin presence [4]. Aplasmomycin is a macrodiolide with a specific inhibitory activity on the futalosine pathway of, for example, *Helicobacter pylori* [5]. Further, it was described to inhibit the growth of several Gram-positive bacteria, including *Mycobacterium* species [6]. Also, 2-alkyl-4-hydroxyquinoline derivatives from *Streptomyces* have bioactive capacities, as some can regulate hyphal growth of *Candida* [7]. These are only a few examples of natural products obtained from marine sources, however, they demonstrate the spectrum of applicability to some extent.

In general, it can be stated that all kinds of bioactivities can be found among the marine-derived natural products, such as those that are antioxidant, anti-Alzheimer, anti-(neuro-)inflammatory, anti-apoptotic, anti-HIV 1, anti-hypertensive, anti-obesity, anti-diabetes, anti-cancer, and radioprotective, as well as those that affect cell-proliferation or show a variety of inhibitory effects.

2. Enzyme Inhibitors from Marine Organisms

Enzyme inhibitors are an important field of research, and they allow us to study enzymes and proteins in more detail, as well as to understand pathways and their regulatory network. Often there is an applied aspect in this research, since many enzyme inhibitor surveys are supposed to identify novel bioactive compounds, precursors, or drugs themselves. These efforts can be tackled via various routes, as, for example, by metabolite screening of known producer organisms while altering growth or environment, direct environmental metabolite screening of rich habitats, screening of related strains or species of known producers, sampling complex cultures or habitats, in silico screening of genomes and metagenomes as well as cloning or manipulating efforts of secondary metabolite gene clusters, docking molecules or libraries into target sites, and undoubtedly more methods to be developed.

A marine *Pseudoalteromonas* strain produces a potent α-d-galactosidase with various fields of application or research interests [8]. Therefore, the identification of potential inhibitory molecules and their mode of action is of importance. Here, the sponge *Monanchora pulchra* was chosen to produce or sample directly candidate compounds such as monanchomycalin B, monanchocidin A, and normonanchocidin A. All three compounds belong to the alkaloid family and are described as pentacyclic guanidine alkaloids [9]. They bound to the target galactosidase and showed an irreversible inhibition. The galactosidase was rescued by the use of D-galactose as a competitive inhibitor. A similar study was done with a α-*N*-acetylgalactosaminidase that showed no inhibitory effects of respective alkaloids, which can be explained by the different protein structure. Therefore, the binding mode of the natural compounds from this sponge is directed by the structure of the target protein. Thus, it can be concluded that the sponge metabolites may have therapeutic effects worth studying and even that those can be directed to specific targets. Furthermore, the two studies herein used α-glycosidases from marine bacteria to study the mode of action of novel alkaloid-like molecules.

Recently, a representative of the *Mycosphaerella* genus was probed for novel secondary metabolites [10], and indeed 11 bioactives were determined and partially characterized from this mangrove fungus. Some structures were verified and, in addition, some biological activities were determined. Three of those, namely asperchalasine, epicoccolide B, and epicolactone, showed an inhibition of α-glucosidase. This is relevant for treating diabetes mellitus (type II) and is of high potential, as limited side effects are expected. Some of the remaining compounds, as well as asperchalasine and epicoccolide B, showed antioxidant activity, which is a beneficial finding and product property.

Marine algae have been used for centuries for various purposes, such as food, cosmetics, or a source of bioactive compounds, of course. It is interesting to see that food-derived molecules can be employed as drugs. For example, the algae *Ulva intestinalis* was used to generate a protein hydrolysate containing potential bioactives [11]. Indeed, a pool of angiotensin I-converting enzyme (ACE) inhibitory peptides was determined and characterized. This provides a simple and non-toxic route towards the development of potential antihypertensive compounds. This is in line with other studies on the isolation of bioactive molecules from marine ecosystems for treatment of hypertension [12,13].

Among the polyphenols, the class of phlorotannins is known to show certain properties of interest for pharma research and chemical biology, for example, the ability to treat diabetes or hypertension as well as to inhibit Alzheimer-related proteins [14]. It was recently shown that the brown seaweed *Ecklonia cava* is a prominent source of such phlorotannins [15,16]. Actually, *E. cava* is edible and thus another food-related source of natural products can be used. This seems to be a growing market, as food may serve in the future as a direct means of access to medicine or therapy, although this needs to be explored in more detail. However, the three phlorotannins eckol, dieckol, and 8,8′-bieckol were reported to exhibit inhibitory effects on Alzheimer-related proteins [14]. In silico studies revealed that the hydroxy groups of these compounds allow interaction with the target enzymes, and thus these are potential drug candidates.

Kinases are often used and are therefore validated targets of therapeutic agents. Therefore, many protein kinase inhibitors have already been reported, but further research is being conducted. However, most of these inhibitory molecules are natural products, and many are of marine origin. The latest findings in this field were recently reviewed and cover novel described compounds from 2014 onwards [17]. It seems as though marine habitats and their organisms bear a number of structural diverse molecules that act as kinase inhibitors. Thus, a number of potential candidates for clinical studies can be derived from marine sources.

Inhibitors also play an important role in diet and the management of nutrition. Notably, in fish farms it is important to keep a well-defined balance of nutrients in order to grow fish properly and keep the population healthy. Thus, (anti-)nutritional factors as part of the supplied feed are described in literature. These substances or their metabolites directly affect growth or health. Those with negative effects are called anti-nutritional factors and are often plant-derived and can represent compounds such as phytohormones, protease inhibitors, and lectins, among others [18]. Little is known about anti-nutritional factors from seaweeds, but these are part of the diet given to fish in fish farming operations. A recent study employed *Ulva ohnoi* as model macroalgae in order to study the effects of protease inhibitors on digestives enzymes of marine fish [18]. It was confirmed that *Ulva* produce protease inhibitors that affect fish. It seems as though these protease inhibitors show a reversible and, notably, mixed-type inhibition towards trypsin and chymotrypsin. These effects can be lowered by a thermal pre-treatment of the algae feed prior to feeding fish. Thus, effects by heat-labile anti-nutritional factors can be overcome. It is necessary to study the structure–function relationship of these algae-derived inhibitors in more detail.

3. Conclusions

The herein given overview of marine habitats, including their organisms, as a potential source of bioactive compounds, and especially enzyme inhibitors, shows that the field is highly dynamic and still growing in many directions. There are still plenty of putative inhibitors and their modes of action to be discovered. Here, novel methods and combinations of established routes will allow further compounds of interest to be uncovered. In particular, the high-throughput sequencing of genomes and metagenomes of even unculturable organisms will allow novel biosynthetic gene clusters to be discovered, which nowadays can be transferred by means of synthetic toolboxes of molecular biotechnology towards producing organisms. Thus, the production of so far silent or uncovered gene clusters will become more and more useful. Screening and robotics will further develop to assist all these efforts. In addition, we now have bioinformatic tools to pursue molecular dynamic studies, including compound docking, and thus targets and mode of action can be predicted with greater and greater accuracy. Therefore, it can be rationalized that we will see more and more of these products from marine environments.

Funding: D.T. was supported by the Federal Ministry for Innovation, Science and Research of North Rhine–Westphalia (PtJ-TRI/1411ng006)—ChemBioCat.

Acknowledgments: D.T. thanks the authors of the articles in this Special Issue for their contributions to the field. Further, the support from the MDPI editorial team is acknowledged.

Conflicts of Interest: The author declares no conflict of interest.

References

1. Narberhaus, F. Blue biotechnology. In *Biotechnology*, 1st ed.; Kück, U., Frankenberg-Dinkel, N., Eds.; De Gruyter: Berlin, Germany, 2015; pp. 123–140.
2. Jang, K.H.; Nam, S.J.; Locke, J.B.; Kauffman, C.A.; Beatty, D.S.; Paul, L.A.; Fenical, W. Anthracimycin, a potent anthrax antibiotic from a marine-derived actinomycete. *Angew. Chem. Int. Ed.* **2013**, *52*, 7822–7824. [CrossRef] [PubMed]
3. Rodríguez, V.; Martín, M.; Sarmiento-Vizcaíno, A.; de la Cruz, M.; García, L.A.; Blanco, G.; Reyes, F. Anthracimycin B, a potent antibiotic against Gram-positive bacteria isolated from cultures of the deep-sea actinomycete *Streptomyces cyaneofuscatus* M-169. *Mar. Drugs* **2018**, *16*, 406. [CrossRef] [PubMed]
4. Hensler, M.E.; Jang, K.H.; Thienphrapa, W.; Vuong, L.; Tran, D.N.; Soubih, E.; Lin, L.; Haste, N.M.; Cunningham, M.L.; Kwan, B.P.; et al. Anthracimycin activity against contemporary methicillin-resistant *Staphylococcus aureus*. *J. Antibiot.* **2014**, *67*, 549–553. [CrossRef] [PubMed]
5. Shimizu, Y.; Ogasawara, Y.; Matsumoto, A.; Dairi, T. Aplasmomycin and boromycin are specific inhibitors of the futalosine pathway. *J. Antibiot.* **2018**, *71*, 968–970. [CrossRef] [PubMed]
6. Okami, Y.; Okazaki, T.; Kitahara, T.; Umezawa, H. Studies on marine microorganisms. V. A new antibiotic, aplasmomycin, produced by a streptomycete isolated from shallow sea mud. *J. Antibiot.* **1976**, *29*, 1019–1025. [CrossRef] [PubMed]
7. Kim, H.; Hwang, J.Y.; Chung, B.; Cho, E.; Bae, S.; Shin, J.; Oh, K.B. 2-Alkyl-4-hydroxyquinolines from a marine-derived *Streptomyces* sp. inhibit hyphal growth induction in *Candida albicans*. *Mar. Drugs* **2019**, *17*, 133. [CrossRef] [PubMed]
8. Bakunina, I.; Likhatskaya, G.; Slepchenko, L.; Balabanova, L.; Tekutyeva, L.; Son, O.; Shubina, L.; Makarieva, T. Effect of pentacyclic guanidine alkaloids from the sponge *Monanchora pulchra* on activity of α-glycosidases from marine bacteria. *Mar. Drugs* **2019**, *17*, 22. [CrossRef] [PubMed]
9. Liu, J.; Li, X.-W.; Guo, Y.-W. Recent advances in the isolation, synthesis and biological activity of marine guanidine alkaloids. *Mar. Drugs* **2017**, *15*, 324. [CrossRef] [PubMed]
10. Qiu, P.; Liu, Z.; Chen, Y.; Cai, R.; Chen, G.; She, Z. Secondary metabolites with α-glucosidase inhibitory activity from the mangrove fungus *Mycosphaerella* sp. SYSU-DZG01. *Mar. Drugs* **2019**, *17*, 483. [CrossRef] [PubMed]
11. Sun, S.; Xu, X.; Sun, X.; Zhang, X.; Chen, X.; Xu, N. Preparation and identification of ACE inhibitory peptides from the marine macroalga *Ulva intestinalis*. *Mar. Drugs* **2019**, *17*, 179. [CrossRef] [PubMed]
12. Tsai, J.-S.; Chen, J.-L.; Pan, B.S. ACE-inhibitory peptides identified from the muscle protein hydrolysate of hard clam (*Meretrix lusoria*). *Process Biochem.* **2008**, *43*, 743–747. [CrossRef]
13. Cao, D.; Lv, X.; Xu, X.; Yu, H.; Sun, X.; Xu, N. Purification and identification of a novel ACE inhibitory peptide from marine alga *Gracilariopsis lemaneiformis* protein hydrolysate. *Eur. Food Res. Technol.* **2017**, *243*, 1829–1837. [CrossRef]
14. Lee, J.; Jun, M. Dual BACE1 and cholinesterase inhibitory effects of phlorotannins from *Ecklonia cava*—An in vitro and in silico study. *Mar. Drugs* **2019**, *17*, 91. [CrossRef] [PubMed]
15. Wijesekara, I.; Yoon, N.; Kim, S. Phlorotannins from *Ecklonia cava* (*Phaeophyceae*): Biological activities and potential health benefits. *Biofactors* **2010**, *36*, 408–414. [CrossRef] [PubMed]
16. Shibata, T.; Kawaguchi, S.; Hama, Y.; Inagaki, M.; Yamaguchi, K.; Nakamura, T. Local and chemical distribution of phlorotannins in brown algae. *J. Appl. Phycol.* **2004**, *16*, 291–296. [CrossRef]
17. Li, T.; Wang, N.; Zhang, T.; Zhang, B.; Sajeevan, T.P.; Joseph, V.; Armstrong, L.; He, S.; Yan, X.; Naman, C.B. A systematic review of recently reported marine derived natural product kinase inhibitors. *Mar. Drugs* **2019**, *17*, 493. [CrossRef] [PubMed]
18. Vizcaíno, A.J.; Galafat, A.; Sáez, M.I.; Martínez, T.F.; Alarcón, F.J. Partial characterization of protease inhibitors of *Ulva ohnoi* and their effect on digestive proteases of marine fish. *Mar. Drugs* **2020**, *18*, 319. [CrossRef] [PubMed]

Article

Preparation and Identification of ACE Inhibitory Peptides from the Marine Macroalga *Ulva intestinalis*

Siqi Sun [1], Xiaoting Xu [1], Xue Sun [1], Xiaoqian Zhang [1], Xinping Chen [2] and Nianjun Xu [1,*]

[1] School of Marine Sciences, Ningbo University, Ningbo 315211, China; 15728041077@163.com (S.S.); m17751232933@163.com (X.X.); sunxue@nbu.edu.cn (X.S.); zhangxiaoqian@nbu.edu.cn (X.Z.)

[2] Division of Allergy, Pulmonary, and Critical Care Medicine, Vanderbilt University Medical Center, Nashville, TN 37232, USA; Xinping.chen@vanderbilt.edu

* Correspondence: xunianjun@nbu.edu.cn

Received: 4 February 2019; Accepted: 15 March 2019; Published: 19 March 2019

Abstract: Angiotensin I-converting enzyme (ACE) inhibitory peptides derived from seaweed represent a potential source of new antihypertensive. The aim of this study was to isolate and purify ACE inhibitory peptides (ACEIPs) from the protein hydrolysate of the marine macroalga *Ulva intestinalis*. *U. intestinalis* protein was hydrolyzed by five different proteases (trypsin, pepsin, papain, α-chymotrypsin, alcalase) to prepare peptides; compared with other hydrolysates, the trypsin hydrolysates exhibited the highest ACE inhibitory activity. The hydrolysis conditions were further optimized by response surface methodology (RSM), and the optimum conditions were as follows: pH 8.4, temperature 28.5 °C, enzyme/protein ratio (E/S) 4.0%, substrate concentration 15 mg/mL, and enzymolysis time 5.0 h. After fractionation and purification by ultrafiltration, gel exclusion chromatography and reverse-phase high-performance liquid chromatography, two novel purified ACE inhibitors with IC_{50} values of 219.35 μM (0.183 mg/mL) and 236.85 μM (0.179 mg/mL) were obtained. The molecular mass and amino acid sequence of the ACE inhibitory peptides were identified as Phe-Gly-Met-Pro-Leu-Asp-Arg (FGMPLDR; MW 834.41 Da) and Met-Glu-Leu-Val-Leu-Arg (MELVLR; MW 759.43 Da) by ultra-performance liquid chromatography-tandem mass spectrometry. A molecular docking study revealed that the ACE inhibitory activities of the peptides were mainly attributable to the hydrogen bond and Zn(II) interactions between the peptides and ACE. The results of this study provide a theoretical basis for the high-valued application of *U. intestinalis* and the development of food-derived ACE inhibitory peptides.

Keywords: *Ulva intestinalis*; ACE inhibitory peptide; optimization; purification; structural identification; molecular docking

1. Introduction

Hypertension, a common, serious chronic disease, affects approximately 25% of the adult population worldwide. Hypertension seriously affects human health and is a causative factor of cardiovascular diseases, stroke, and renal diseases, among others [1,2]. The renin–angiotensin system (RAS) and kallikrein–kinin system (KKS) are crucial for regulating blood pressure in the human body. Angiotensin I-converting enzyme (E.C.3.4.15.1, ACE), a peptidase belonging to the zinc metalloenzyme family, plays an important role in RAS and KKS, inactivating angiotensin I to the potent vasoconstrictor angiotensin II and also inactivating the vasodilator bradykinin to raise blood pressure [3]. Therefore, inhibition of ACE activity is effective for maintaining blood pressure within a normal range [4].

ACE inhibitors (ACEIs) inhibit ACE activity and reduce blood pressure by inhibiting the synthesis of angiotensin II or promoting the release of bradykinin. Although ACEIs, such as enalapril, captopril, and lisinopril, are widely used in hypertension treatments, synthetic ACEIs have a series of negative effects, including hypotension, cough, increased potassium levels and, angioedema [5,6]. Consequently, the development of safe and effective antihypertensive drugs is important, and due to antihypertensive effects and safety, there has been an increasing interest in food-derived ACE inhibitory peptides (ACEIPs) during the last decades. To date, ACEIPs derived from a variety of products such as milk [7], bovine collagen [8], mushrooms [9], rice [10], and marine sources including fish, shellfish, and macroalgae [11,12], have been reported. Indeed, marine organisms, which are rich in unique bioactive compounds, are valuable for human health. Hence, there is much research attention in isolating bioactive compounds from marine organisms to develop new drugs or health products. Macroalgae are important bio-resource organisms in marine ecosystems. According to previous studies, many unique bioactive compounds, including peptides, fats, and carbohydrates, have been isolated from macroalgae [13,14]. Moreover, some novel ACE inhibitory peptides with efficient antihypertensive effects have been isolated from enzymatic hydrolysates of algal species. For instance, Cao et al. reported a peptide with an IC_{50} value of 474.36 μM (Gln-Val-Glu-Tyr) from hydrolyzed *Gracilariopsis lemaneiformis* [12]. Other peptides, such as Ala-Ile-Tyr-Lys (IC_{50} = 213 μM), Tyr-Lys-Tyr-Tyr (IC_{50} = 64.2 μM), Lys-Phe-Tyr-Gly (IC_{50} = 90.5 μM), and Tyr-Asn-Lys-Leu (IC_{50} = 21 μM) from *Undaria pinnatifida* [15] and Ile-Tyr (IC_{50} = 2.69 μM), Ala-Lys-Tyr-Ser-Tyr (IC_{50} = 1.52 μM), Leu-Arg-Tyr (IC_{50} = 5.06 μM), and Met-Lys-Tyr (IC_{50} = 7.26 μM) from *Porphyra yezoensis* [16] have been found. In addition, two peptides (Ile-Pro and Ala-Phe-Leu) with IC_{50} values of 87.6 μM and 65.8 μM were purified from an *Ulva rigida* protein hydrolysate [17]. Some peptides have exhibited powerful antihypertensive effects comparable to those of pharmaceutical drugs in spontaneously hypertensive rats (SHRs) [16,18]. Thus, marine algae can be used as a new source of ACEIPs.

Ulva intestinalis, a marine green algae belonging to the family of Ulvaceae, consisting of a tubular frond and unbranched thalli [19]. It is able to reproduce using unfused gametes, spores, and zygotes. Under suitable growth conditions *U. intestinalis* can quickly occupy the littoral zone [20], and it is among the species that cause green tides, which can affect the growth of other coastal organisms [21,22]. In addition, *U. intestinalis* is regularly consumed in the East Asian countries of China, Korea, and Japan. It has been reported that *U. intestinalis* is rich in vitamins (0.174 mg/g), proteins (~20.5%), carbohydrates (42.1%), and other bioactive compounds [23,24], and the high content of crude proteins in *U. intestinalis* renders it a potential source of ACE inhibitory peptides for the functional foods and medical industries. To the best of our knowledge, no study to date has aimed at purifying and characterizing ACE inhibitory peptides from *U. intestinalis*.

In this study, *U. intestinalis* was hydrolyzed using five different proteases (trypsin, pepsin, papain, α-chymotrypsin, alcalase), and response surface methodology (RSM) was employed to optimize the hydrolysis conditions, including pH, hydrolysis temperature, substrate concentration, and enzyme/substrate ratio (E/S). The hydrolysate solution was fractionated using nominal molecular weight limit (NMWL) Amicon Ultra-15 centrifugal filters, and bioactive peptides were further purified and identified using Sephadex G-25, G-15 gel chromatography, reverse-phase high-performance liquid chromatography (RP-HPLC), and ultra-performance liquid chromatography-tandem mass spectrometry (UPLC-MS/MS). Furthermore, the peptides were chemically synthesized and then used for the determination of their stability during gastrointestinal digestion.

2. Results and Discussion

2.1. Preparation of ACE Inhibitory Peptides from U. intestinalis

Proteases are necessary to release ACEIPs from inactive forms [25]. Different proteases affect the composition and size of the polypeptides produced, which can affect their biological activities [26]. In this study, *U. intestinalis* proteins were hydrolyzed using five different proteases, and their ACE inhibitory activities were assessed (Figure 1). The trypsin-hydrolyzed product showed the greatest ACE inhibitory activity (51.15 ± 3.78%). Thus, trypsin was chosen for the production of ACEIPs.

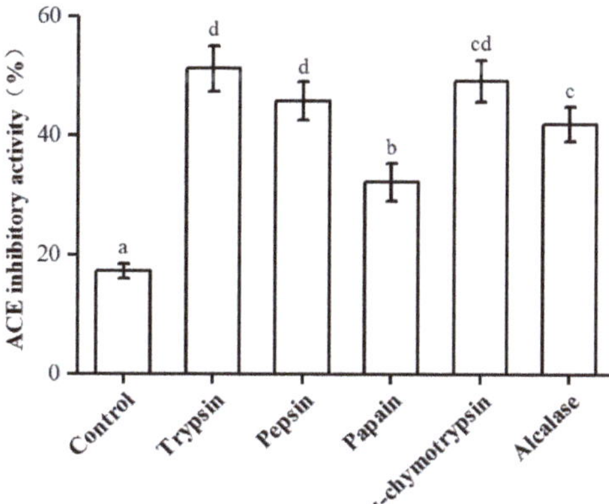

Figure 1. Angiotensin I-converting enzyme (ACE) inhibitory activities of *U. intestinalis* protein hydrolysates produced by different enzymes. Each point is the mean of three determinations ($n = 3$) ± SD. Different letters indicate significant differences. The concentration of each hydrolysate was 2.5 mg/mL; crude protein (2.5 mg/mL) was used as the positive control.

Various factors such as temperature, pH, and substrate concentration affect protein extraction from marine sources [27], and the effect of pH on the ACE inhibition rate of the hydrolysates was significant, as shown in Figure 2A. With a rise in pH, the ACE inhibitory activity of the protein hydrolysates increased to a maximum value (60.98%) at pH 8.0; there was no significant difference in ACE inhibitory activity at a higher pH. A certain range of pH can affect the degree of dissociation of enzyme molecules and substrates, and promote the binding of enzymes to substrates. Therefore, the primary enzyme solution pH is 8.0.

Temperature is also an important factor for influencing enzyme activity and hydrolysis efficiency. Figure 2B shows the effect of temperature on the ACE inhibitory activity of protein hydrolysates. ACE inhibitory activity gradually increased from 22 to 32 °C, and the maximum value was reached at 32 °C (59.53%), above which the inhibitory activity decreased. The reason for this may be that trypsin is gradually activated as the temperature of enzymatic hydrolysis increases, resulting in more active peptide fragments, and the structure of the enzyme protein was affected when the temperature exceeded the optimum temperature. Therefore, the optimal temperature for the production of ACEIPs from *U. intestinalis* was determined to be 32 °C (Figure 2B).

The effect of substrate concentration from 5 to 30 mg/mL on the ACE inhibitory activity of *U. intestinalis* protein hydrolysates was also evaluated. As presented in Figure 2C, inhibitory activity increased from 5 to 10 mg/mL, but rapidly decreased from 15 to 30 mg/mL. This may have been due to the high concentration of substrate affecting the binding of the protease to the substrate, thereby inhibiting the enzymatic reaction. The highest value of ACE inhibitory activity (57.19%) was observed at a substrate concentration of 15 mg/mL, which was chosen for further experiments.

The effect of E/S on the ACE inhibitory activity of the hydrolysates was also studied. At a range from 1 to 6%, the ACE inhibitory activity reached a maximum (60.94%) at an E/S of 4% (Figure 2D). Thus, 4%was considered to be the optimal E/S in this study. The possible explanation for this is that an increase in the E/S, which increases the chance of contact between the substrate protein and the enzyme, accelerates the enzymatic reaction and produces more active peptides.

As depicted in Figure 2E, no significant difference in ACE inhibitory activity was observed when the reaction time increased from 2 to 8 h. In the interest of time concerns, 2 h was chosen as the hydrolysis time for ensuing experiments.

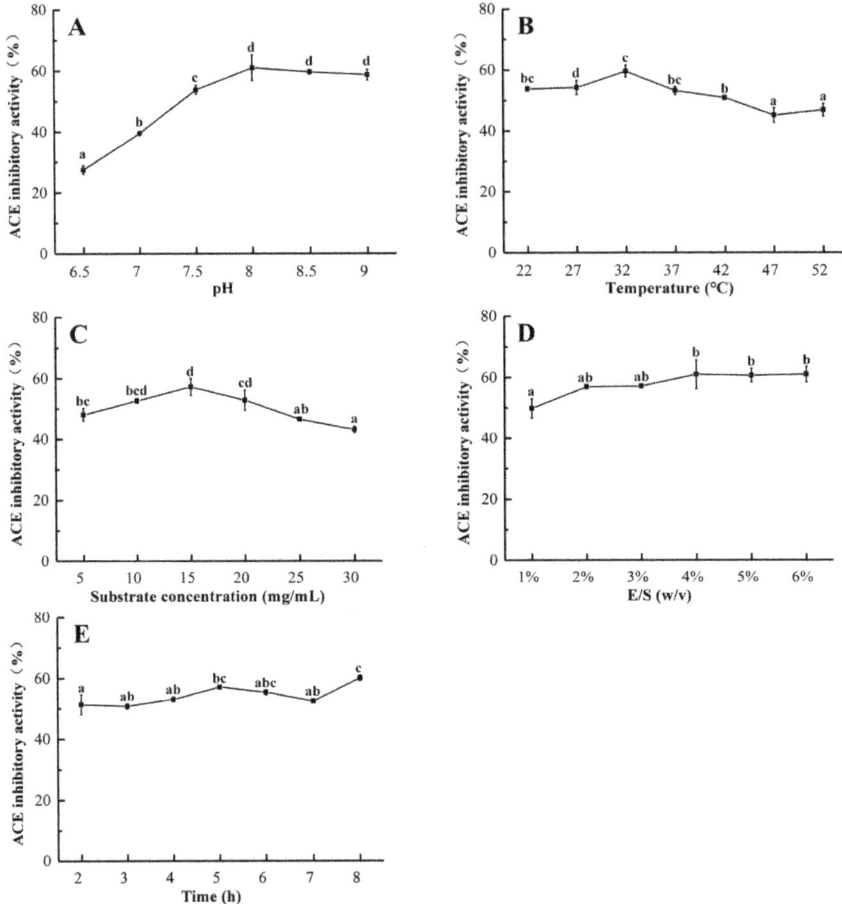

Figure 2. Effects of hydrolysis pH (**A**), temperature (**B**), substrate concentration (**C**), E/S ratio (**D**), and reaction time (**E**) on ACE inhibitory activity of protein hydrolysates from *U. intestinalis*.

2.2. Optimization of the Enzymatic Hydrolysis Condition

RSM was utilized to optimize the enzymatic hydrolysis conditions with regard to three significant factors for the production of ACE inhibitory peptides, including pH (*X*1), temperature (*X*2), and substrate concentration (*X*3). The factors and levels are provided in Table 1. The experimental design and Box–Behnken results for the incubation conditions are shown in Table 2, where *Y* represents the ACE inhibitory activity and *X*1, *X*2, and *X*3 represent the pH, temperature, and substrate concentration, respectively.

Table 1. Coded values and independent variables of the response surface experiment.

Coded Level	Independent Variable		
	***X*1: pH**	***X*2: Temperature (°C)**	***X*3: Substrate Concentration (mg/mL)**
−1	7.5	27	10
0	8.0	32	15
1	8.5	37	20

Table 2. Coded values and independent variables of the response surface experiment.

No	*X1*	*X2*	*X3*	ACE Inhibition (%)
1	−1	−1	0	50.43
2	1	−1	0	62.61
3	−1	1	0	55.36
4	1	1	0	54.78
5	−1	0	−1	56.23
6	1	0	−1	57.39
7	−1	0	1	52.61
8	1	0	1	63.91
9	0	−1	−1	60.58
10	0	1	−1	59.42
11	0	−1	1	60.00
12	0	1	1	59.13
13	0	0	0	61.74
14	0	0	0	61.74
15	0	0	0	63.04
16	0	0	0	63.48
17	0	0	0	63.48

The data were analyzed to obtain a quadratic regression model using Design Export 8.0.6 (StatEase, Inc, USA). A multiple regression equation correlating to the response function with independent variables was as follows:

$$Y = 62.70 + 3.01A - 0.62B + 0.25C - 3.19AB + 2.54AC + 0.072BC - 4.57A^2 - 2.33B^2 - 0.59C^2 \quad (1)$$

The results of analysis of variance and the fitness of the model are summarized in Table 3. Overall, pH, pH and temperature, and pH and substrate concentration had significant effects on inhibitory activities ($p < 0.001$). As the value of the "lack of fit" item was $p = 0.5449$, unknown factors had little influence on the results.

Table 3. Variance analysis for ACE inhibitory activity in the RSM test.

Source	Sum of Squares	Df	Mean Square	*F* Value	*p* Value	Significance
Model	262.15	9	29.13	39.67	<0.0001	**
A-pH	72.35	1	72.35	98.53	<0.0001	**
B-Temperature	3.04	1	3.04	4.13	0.0815	
C-Substrate concentration	0.51	1	0.51	0.7	0.4301	
AB	40.66	1	40.66	55.38	0.0001	**
AC	25.73	1	25.73	35.04	0.0006	**
BC	0.021	1	0.021	0.029	0.8705	
A^2	88.03	1	88.03	119.89	<0.0001	**
B^2	22.78	1	22.78	31.03	0.0008	**
C^2	1.45	1	1.45	1.98	0.2026	
Residual	5.14	7	0.73			
Lack of Fit	1.96	3	0.65	0.82	0.5449	
Pure Error	3.18	4	0.79			
Corrected Total	267.29	16				

Df: degrees of freedom, MS: mean square, *F* and *p* values, respectively ** $p < 0.001$, extremely significant.

As indicated in Table 3, the regression model was also used to fit the effect of three factors on the ACE inhibition rate. The coefficient of multiple determinations (R^2) for the quadratic regression model was 0.9808; thus, a 98.08% response to the ACE inhibition rate was caused by the concentration of A, B, and C and their interactions. Moreover, the value of adjusted determination coefficients (R^2adj) and R^2

were both close to 1 (R^2 = 0.9808 and R^2_{adj} = 0.9560), indicating that this model may be used to analyze and predict changes in ACE inhibitory activity under different enzymatic hydrolysis conditions [28].

In our present study, response surface plots and contour plots were applied to demonstrate the effect and interaction of independent variables on the ACE inhibitory rates of protein hydrolysates. Figure 3 illustrates the effect of X1 and X2 on such ACE inhibitory activities.

The optimum condition for ACE inhibitory activity was obtained at pH 8.42, 28.5 °C, a substrate concentration of 15 mg/mL, an E/S of 4%, and an enzymolysis time of 5 h. Under optimal reaction conditions, the predicted and experimental values for the ACE inhibitory activities of the protein hydrolysates were 64.91 and 64.07%, respectively, indicating that the predicted value was close to the experimental value. Thus, the parameters obtained by the RSM optimizations were reliable, and it is feasible to use them in practice.

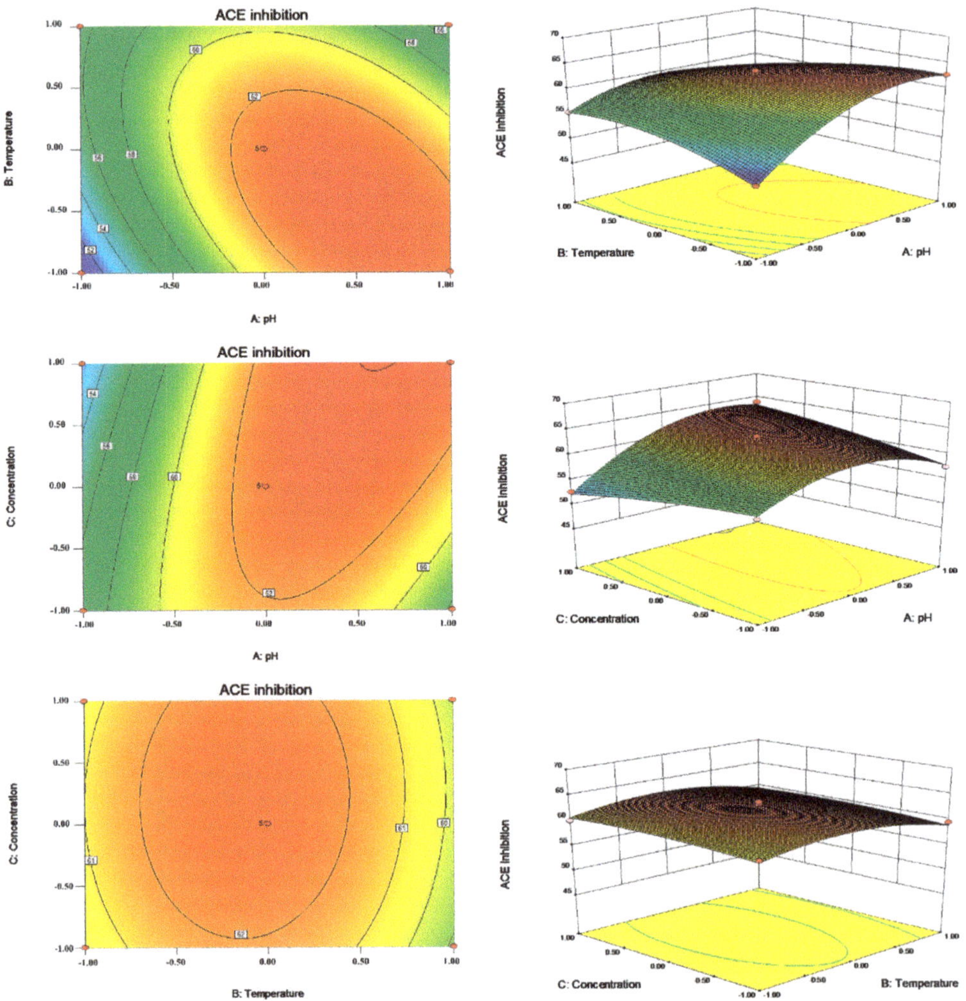

Figure 3. Contour plots and response surface plots for pH (**A**), temperature (**B**), and substrate concentration (**C**) to ACE inhibition rate.

2.3. Purification of ACE Inhibitory Peptides

Ultrafiltration can separate protein hydrolysates into components of different molecular weights (MWs), and active peptides with different components have different biological activities. It has been reported that peptides with molecular weights <3 kDa generally possess high ACE inhibitory activity [29,30]. In the present study, the hydrolysate was separated into three fractions (<3 kDa, 3–10 kDa, and >10 kDa) by filtering with ultrafiltration membranes. The IC_{50} value and ACE inhibitory activity at 1.5 mg/mL of each fraction were assessed, and the results are displayed in Table 4. Among the fractions, the MW < 3 kDa fraction exhibited the strongest ACE inhibitory activity, with an inhibitory rate of 53.01%. In contrast, peptides with a high molecular weight (MW > 10 kDa) showed lower ACE inhibitory activity (Table 4). Thus, the fraction with MW < 3 kDa was chosen for further separation and purification of ACEIPs.

Table 4. The ACE inhibitory of activity of the fraction separated by ultra-filtration.

Fraction	IC_{50} (mg/mL)	ACE Inhibitory Activity (%) 1.5 mg/mL
Unfractionated	1.59 ± 0.08 [a]	48.72 ± 1.13 [a]
MW < 3 kDa	1.14 ± 0.11 [b]	53.01 ± 0.85 [a]
3 kDa < MW < 10 kDa	2.19 ± 0.08 [c]	44.16 ± 0.85 [b]
MW > 10 kDa	2.53 ± 0.17 [d]	34.76 ± 0.85 [c]

Values are presented as mean ± standard deviations from triplicates ($n = 3$). Means with different lower case letters are significantly different ($p < 0.05$).

The MW < 3 kDa fraction was further separated using a Sephadex G-25 gel filtration column, with five major peaks at 220 nm (Figure 4A): A, B, C, D, and E. At a concentration of 1.5 mg/mL, fractions B, C, and D showed inhibitory activity against ACE, with minimal activity for fraction A. With an inhibitory rate of 56.3%, fraction C exhibited the greatest ACE inhibitory activity (Figure 4B).

Accordingly, fraction C was further separated using a Sephadex G-15 gel filtration column, revealing three major peaks (C1–3) at 220 nm (Figure 4C). Among them, fraction C2 showed the highest activity (Figure 4D).

Fraction C2 was then concentrated and used for further separation by means of RP-HPLC. The solution was purified on an AKTA pure system (GE Healthcare, Uppsala, Sweden) with an Inertsil ODS-3 C18 column ($\varphi 10 \times 250$ mm). Nine peaks were collected separately (Figure 5A). Among those fractions, C2-8 exhibited the most potent ACE inhibitory activity, with an inhibitory rate of 62.35% (Figure 5B). Then, the fraction C2-8 was further purified by HPLC.

As shown in Table 5, two ACEIPs were obtained from fraction C2-8 by UPLC-MS/MS and de novo sequencing. According to mass spectra determined by UPLC-MS/MS (Figure 6), Mascot software identified Phe-Gly-Met-Pro-Leu-Asp-Arg and Met-Glu-Leu-Val-Leu-Arg (FGMPLDR and MELVLR; Matrix Science, Inc, USA), which are novel peptides with ACE inhibitory activity from *U. intestinalis* activity not previously reported. The IC_{50} values of FGMPLDR and MELVLR were 219.35 μM and 236.85 μM, respectively. Previous studies have shown that the amino acid composition of a peptide has a significant effect on its ACE inhibitory activity [31], and it has been reported that ACE prefers to bind to a polypeptide with a high content of hydrophobic amino acids [32]. The two peptides obtained in this experiment both have a hydrophobic amino acid content, 42.9 and 50%, for FGMPLDR and MELVLR, respectively, which may contribute to their high activities. Furthermore, FGMPLDR and MELVLR are composed of 6–7 amino acids, which may also play a crucial role in their ACE inhibitory activities. According to previous studies, most ACEIPs are small peptides of 2–12 residues and molecular weights less than 3000 Da [33], which may more easily fit in the ACE active site and thus assert inhibitory activity [34]. The composition of the C- and N-terminal residues of an ACEIP also have a significant effect on ACE inhibition rate, with high activity when the C-terminal residue is Tyr, Phe, Pro, Trp, or Leu and the N-terminal residue is a hydrophobic aliphatic branched-chain amino acid such as Leu, Ile, Ala, or Met. Moreover, positively charged amino acids such as Arg and Lys at the C-terminus and a basic amino acid (Arg, Lys, and His) at the N-terminus can enhance the affinity of the peptide for ACE, further increasing antihypertensive activity [35–37]. The C-terminal amino acids of the two peptides in our study (FGMPLDR and MELVLR) were Arg, which was consistent with the results of a previous study [36].

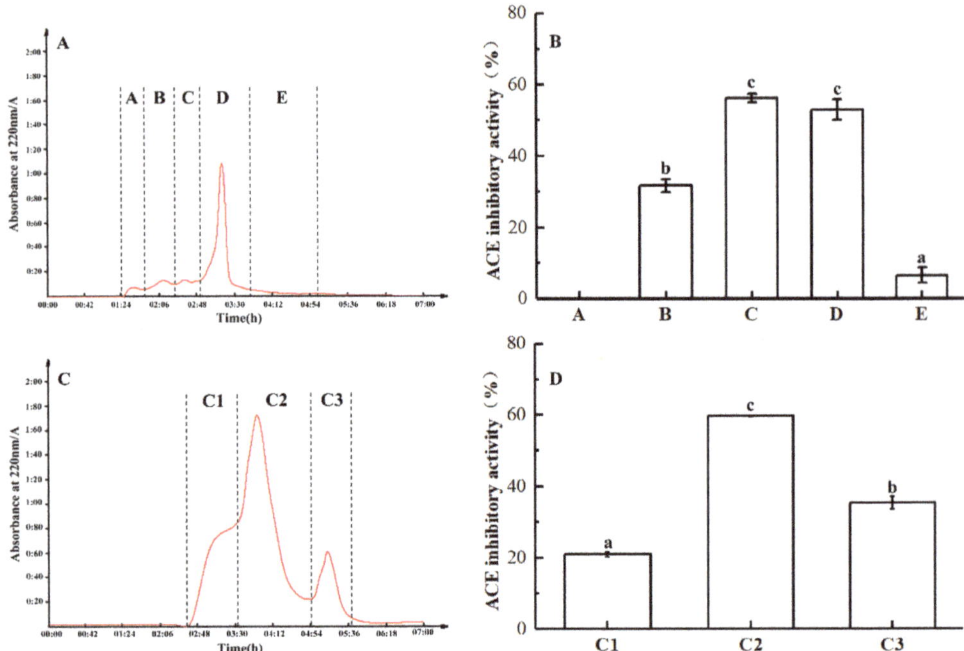

Figure 4. Sephadex G-25 gel filtration chromatogram of <3 kDa fraction of trypsin hydrolysate from *U. intestinalis*. (**A**) The fraction was divided into five parts (**A–E**) by Sephadex G-25. (**B**) The ACE inhibitory activity (1.5 mg/mL) and percentage of **A–E**. (**C**) The fraction was divided into three parts (C1–C3) by Sephadex G-15. (**D**) The ACE inhibitory activity (1.5 mg/mL) and percentage of C1 to C3.

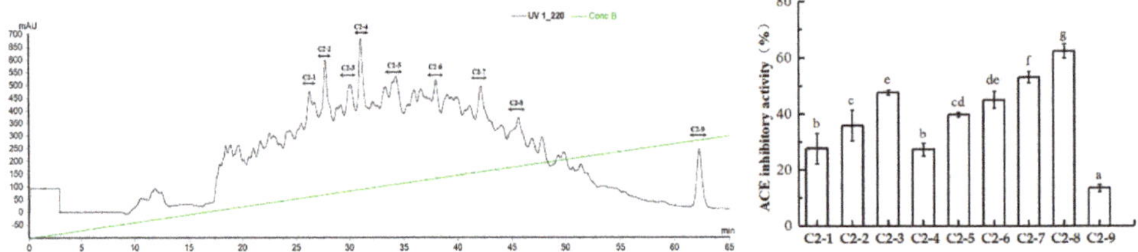

Figure 5. ACE inhibitory activity of fraction C2 from RP-HPLC. (**A**) The fraction was divided into 9 parts (C2-1 to C2-9) by RP-HPLC. (**B**) ACE inhibitory activity (1.5 mg/mL) and percentages of C2-1 to C2-9. Means with different lower case letters are significantly different ($p < 0.05$).

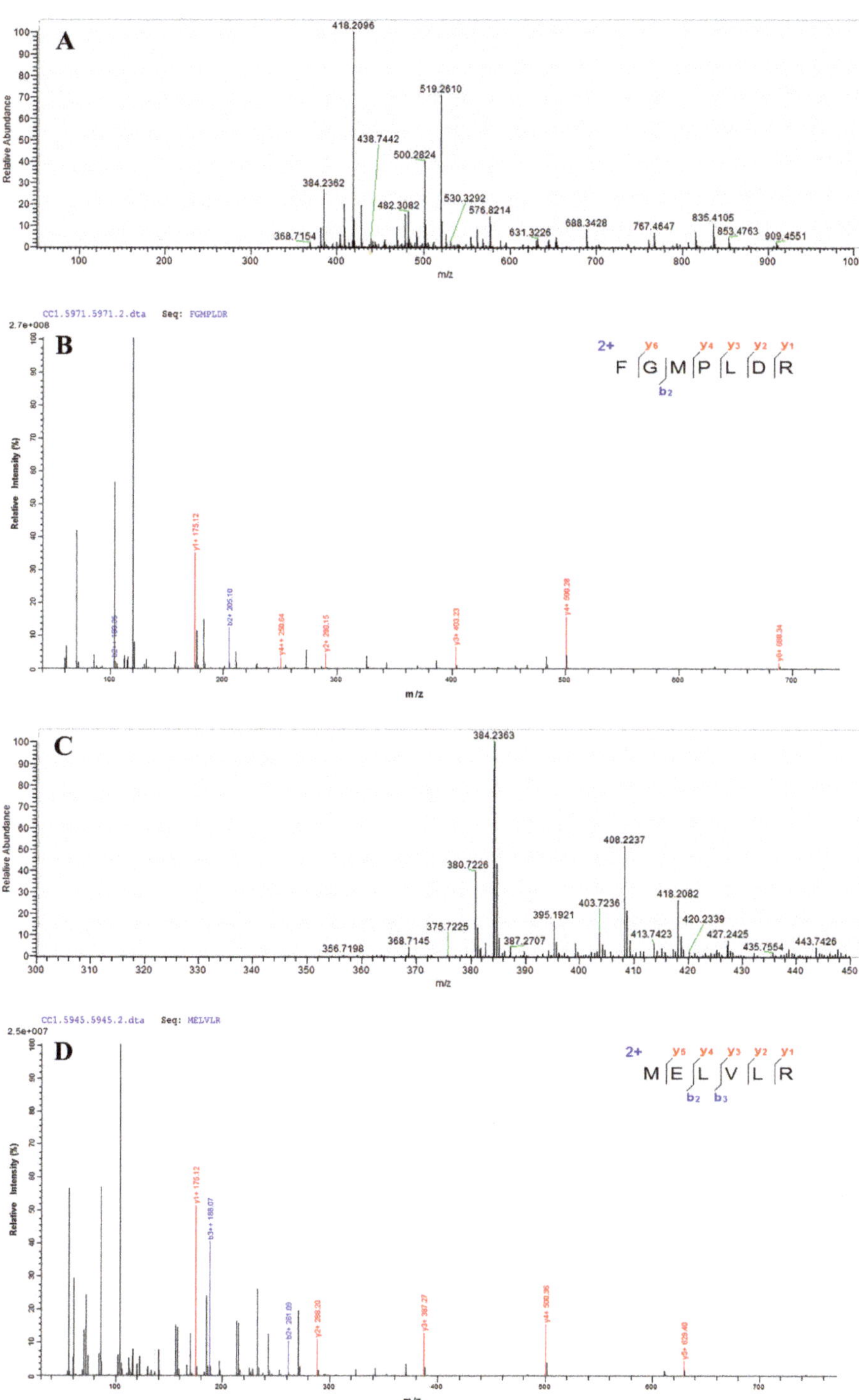

Figure 6. The primary mass spectrogram and corresponding secondary mass spectrogram of FGMPLDR and MELVLR. (**A**) MS/MS spectra of FGMPLDR. (**B**) The secondary mass spectrogram of FGMPLDR. (**C**) MS/MS spectra of MELVLR. (**D**) The secondary mass spectrogram of MELVLR.

Table 5. Peptides identified in fraction C2-8.

Amino Acid Sequence Analysis	Mass	*m*/*z*	*z*	Area	ALC (%)	IC_{50} (μM)
FGMPLDR	834.41	418.2096	2	1.69×10^9	99	219.35
MELVLR	759.43	380.7226	2	3.48×10^8	99	236.85

2.4. In Vitro Stability of ACEIPs Derived from *U. intestinalis*

After gastrointestinal digestion, some food-derived ACEIPs do not exhibit (or exhibit fewer than) the expected hypotensive effects [38,39]. Thus, to evaluate resistance to gastrointestinal enzymes, the ACEIPs obtained in our study were subjected to a two-step hydrolysis process. After digestion with pepsin and trypsin, the ACE inhibitory activities of FGMPLDR and MELVLR were 51.32 and 58.63%, respectively, with no significant difference from the control (Table 6). Therefore, our results indicate that these peptides are stable in the gastrointestinal tract and may also show effective antihypertensive activity in vivo.

Table 6. Simulated gastrointestinal digestion of synthetic peptides at 0.2 mg/mL.

Enzyme	ACE Inhibitory Activity (%)	
	FGMPLDR	MELVLR
Control	50.07 ± 0.07	59.92 ± 0.04
Pesin [a]	52.98 ± 0.19	56.84 ± 0.04
Pesin + Trypsin [b]	51.32 ± 0.02	58.63 ± 0.13

[a] Pepsin hydrolysis for 2 h; [b] Pepsin hydrolysis for 2 h followed by trypsin hydrolysis for 2 h.

2.5. Molecular Docking

To elucidate the inhibitory mechanism, docking simulation was conducted using AutoDock 4.2 software. The best results were obtained for FGMPLDR and MELVLR at the ACE active site in the presence of Zn(II) (Figure 7), with binding energies of −2.78 kcal/mol and −6.04 kcal/mol, respectively. This is a reference for assessing the binding between proteins and peptides [38,40]. The peptides and ACE residues are mainly linked through hydrogen bonds, hydrophobic interaction, and polar, Van der Waals, and electrostatic forces. It has been reported that hydrogen bond interactions play an irreplaceable role in stabilizing the structure of the complex as well as the ACE reaction [41,42]. Previous studies have indicated three main active site pockets in the ACE molecule. The S1 pocket includes three residues, Ala354, Glu384, and Tyr523, and the S2 pocket Gln281, His353, Lys511, His513, and Tyr520; in contrast, the S1′ pocket only includes residue Glu162 [43]. Furthermore, the lisinopril, an ACE inhibitor, was found to share interactions at Ala354, His383, Glu384, and Lys511, showing that those residues might play major roles in ACE binding [44,45]. Our molecular docking studies indicated that FGMPLDR and MELVLR bind to the active site pocket of ACE through a network of hydrogen bonds and hydrophobic and Van der Waals interactions. Both peptides displayed a stable docking structure with ACE. As shown in Figure 7B, five hydrogen bonds between FGMPLDR and residues Glu123, Ala354, Ala356, Glu384, and Arg522 of ACE were formed, and the Van der Waals forces for nine residues were also important. Namely, two hydrogen bonds with the S1 pocket (Ala354 and Glu384) were produced for FGMPLDR. In the case of MELVLR, it formed six hydrogen bonds with residues Asn70, Glu143, Gln281, His383, and Lys511, and hydrophobic interactions with seven residues (Figure 7D). Gln281 and Lys511 of the S2 pocket associated with MELVLR through two hydrogen bonds. Furthermore, as ACE is a metalloenzyme with a zinc ion coordinated in the active site with His348, Glu372, and His344 [38], the presence of Zn(II) plays an important role in ACE inhibition [46]. For peptides FGMPLDR and MELVLR, Gly and Leu are coordinated to the Zn(II) ion, respectively.

This interaction may have caused distortion of the tetrahedrally coordinated Zn(II), which further resulted in the loss of ACE activity.

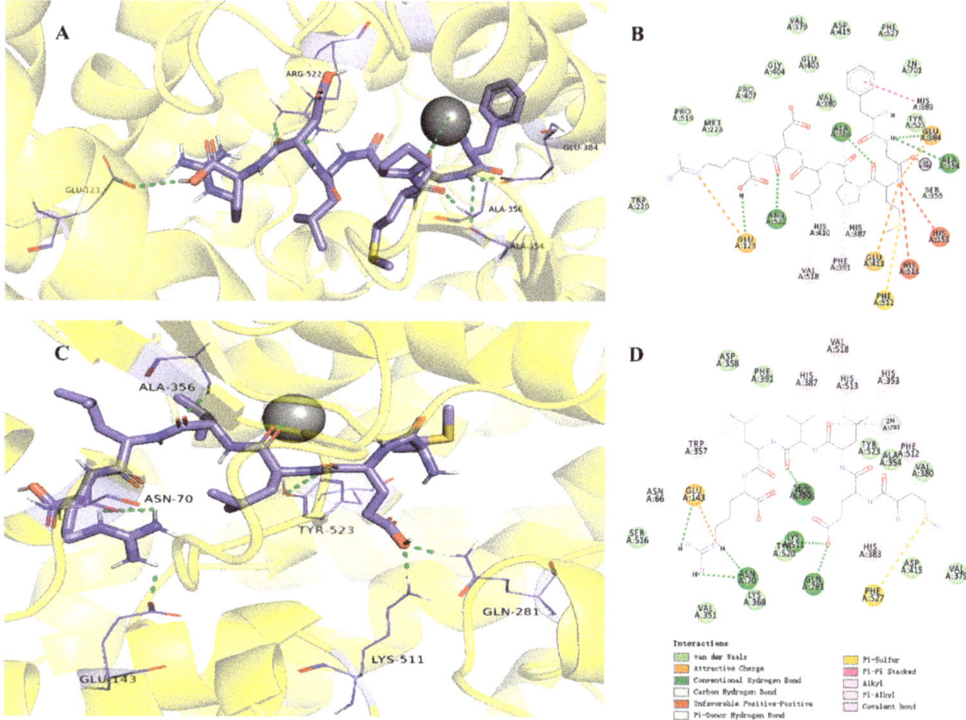

Figure 7. Molecular docking results for FGMPLDR and MELVLR with ACE (PDB: 1O8A). (**A**) 3-D details of ACE and FGMPLDR interactions. (**B**) 2-D interaction details for FGMPLDR. (**C**) 3-D details of ACE and MELVLR interactions. (**D**) 2D interaction details for MELVLR.

3. Materials and Methods

3.1. Materials and Chemicals

Fresh *U. intestinalis* was collected from *Porphyra yezoensis* aquaculture rafts (N 29°44′, S 121°54′). The samples were washed with sterile water twice to remove any adherents and necrotic parts and then dried on paper.

Trypsin, pepsin, papain, α-chymotrypsin, alcalase, and 3.5-kDa dialysis tubing were purchased from Solarbio Science and Technology Co., Ltd. (Beijing, China). ACE (from rabbit lung), *N*-[3-(2-Furyl)acryloyl]-Phe-Gly-Gly (FAPGG), Sephadex G-15, Sephadex G-25, and acetonitrile (ACN) were obtained from Sigma Chemical Co. (St. Louis, MO, USA). The 10-kDa and 3-kDa nominal molecular weight limit (NMWL) Amicon Ultra-15 centrifugal filters were purchased from Merck Millipore (Darmstadt, Germany). Other chemicals and reagents were of analytical grade.

3.2. Preparation of *U. intestinalis* Protein

Freeze-thaw with the frequency ultrasonic method [47] was used to extract protein from *U. intestinalis*. Samples were ground with three volumes of 2% NaCl. The mixture was frozen at −40 °C for 6 h and then incubated in a water bath at 20 °C to thaw (repeated three times). The mixture was further extracted using an ultrasonic processor (Scientz-IID; Scientz, Ningbo, China) at 25 kHz at 300 W for 25 min, and the homogenates were centrifuged (Sorvall ST 16R Centrifuge, Thermo Electron LED GmbH, Osterode, Germany) with 8000 rpm for 15 min at 4 °C. Ammonium sulfate was added to the supernatant at a concentration of 60% to precipitate protein, which was collected by concentration (4 °C, 8000 rpm, 15 min). The protein pellet was dissolved in distilled water and dialyzed for 48 h at 4 °C using a 3.5-kDa MWCO dialysis bag against distilled water; the dialyzed retentate was lyophilized.

3.3. Enzymatic Hydrolysis of U. intestinalis Protein

U. intestinalis protein was digested with trypsin (37 °C, pH 8.0), pepsin (37 °C, pH 2.0), papain (37 °C, pH 6.0), α-chymotrypsin (37 °C, pH 8.0), and alcalase (37 °C, pH 10.0) for 5 h.The substrate concentration and enzyme/protein ratio were fixed at 20.0 mg/mL and 4% (*w*/*w*), respectively. The reaction was stopped by heating at 100 °C for 15 min, and the protein hydrolysates were centrifuged with 8000 rpm at 4 °C for 15 min. The supernatants were lyophilized and stored at −20 °C until use.

3.4. Single-Factor Experimental Design

Single-factor experiments were designed to obtain relevant factors for the production of ACEIPs and the experimental ranges of RSM. Based on previous experiments, trypsin was chosen as the optimal enzyme. The digestion conditions for the single-factor experiments, including pH (6.5, 7.0, 7.5, 8.0, 8.5, and 9.0), temperature (22, 27, 32, 37, 42, and 47 °C), substrate concentration (5.0, 10.0, 15.0, 20.0, 25.0, 30.0, and 35.0 mg/mL), E/S (1.0, 2.0, 3.0, 4.0, 5.0, and 6.0%), and reaction time (2.0, 3.0, 4.0, 5.0, 6.0, and 7.0 h), were investigated to reveal their influences on the ACE inhibitory activity of the protein hydrolysates.

3.5. RSM Experimental Design

In combination with the results obtained from the single-factor experiment, RSM was applied to optimize the enzymatic hydrolysis conditions. E/S and the reaction time were kept constant, at 4.0% and 5 h, respectively. The independent variables included pH ($X1$), temperature ($X2$), and substrate concentration ($X3$); the response variable (Y) was ACE inhibitory activity. The processing parameters were optimized using a Box–Behnken design, and each selected variable was coded at three levels (−1, 0, +1).

The software Design-Expert 8.0.6 was used to perform the experimental design and regression analysis of the experimental data. For verification of the predictive enzymatic hydrolysis conditions model, we further determined the ACE inhibitory activity of the enzymatic hydrolysate produced under optimum conditions.

3.6. ACE Inhibition and IC_{50} Assay

ACE inhibition was examined according to Shalaby et al. and Henda et al. [48,49], with slight modifications. FAPGG and ACE were dissolved in 50 mM Tris-HCl (pH 7.5) containing 300 mM NaCl. A sample solution (40 μL) was mixed with 100 μL of 0.88 mM FAPGG solution, and the mixture was incubated at 37 °C for 5 min. To start the reaction, 60 μL of ACE solution (0.20 U/mL) was added; absorbance was measured at 340 nm and recorded every 1 min for 30 min using a SpectraMax 190 absorbance microplate reader (Molecular Devices, Sunnyvale, USA). The slope averaged over a linear interval of 10–30 min was taken as a measure of the ACE inhibitory activity. The degree of ACE inhibition was calculated according to the following equation:

$$\text{ACE inhibition rate (\%)} = [1 - (\text{slope inhibitior}/\text{slope control})] \times 100 \quad (2)$$

The activity of each sample was tested in triplicate. The IC_{50} value, the concentration of peptide required to reduce ACE activity by 50%, was determined by regression analysis of ACE inhibition (%) vs. peptide concentration.

3.7. Separation and Purification of ACE Inhibitory Peptide

3.7.1. Ultrafiltration Separation

The hydrolysate solution was passed through ultrafiltration membranes with molecular weight (MW) cutoffs of 10 and 3 kDa to obtain three fractions (MW < 3 kDa; MW 3–10 kDa; MW > 10 kDa). Each fraction was assessed for ACE inhibitory activity and IC_{50}.

3.7.2. Gel Filtration Chromatography Analysis

The fraction with the highest ACE inhibition rate was applied to a Sephadex G-25 column (φ2.5 × 60 cm) at a flow rate of 0.5 mL/min using an MC99-3 automatic liquid chromatograph (Shanghai Huxi Analysis Instrument Factory Co., Ltd., Shanghai, China). Elution was measured at 220 nm, and 3-mL fractions were collected. Each major peak was pooled and lyophilized for ACE inhibitory activity assays. Fraction C with the highest activity was further fractionated using a Sephadex G-15 gel filtration column (φ1.0 × 60 cm) with the same operation as described above for Sephadex G-25.

3.7.3. RP-HPLC Analysis of ACE Inhibitory Peptides

The selected fraction C2 was fractionated using semi-preparatory RP-HPLC. Briefly, 7.5 mg of the peptide fraction was dissolved in 1 mL 0.05% TFA (*v*/*v*) and passed through a 0.2-μm filter. The solution was purified using an AKTA pure system (GE Healthcare, Uppsala, Sweden) with an Inertsil ODS-3 C18 column (φ10 × 250 mm). Ultra-pure water containing 0.1% TFA (*v*/*v*) was used as mobile phase A, and acetonitrile containing 0.1% TFA (*v*/*v*) was used as mobile phase B. Separation of the peptides was carried out at a flow rate of 1.5 mL/min using the following linear gradient: 0 to 50% eluent B from 0 to 65 min. Elution peaks were monitored at 220 nm. The purity of the fraction with the highest activity was further analyzed using a ZORBAX Eclipse XDB-C18 column (5 μm, φ4.6 mm × 150 mm; Agilent, USA). The column was eluted with a linear gradient of 0–35% mobile phase B from 0 to 25 min at a flow rate of 1 mL/min. Collected peaks were concentrated using a rotary evaporator system and lyophilized for later use.

3.8. Identification of the Amino Acid Sequence by UPLC-MS/MS

Identification of amino acid sequences was achieved using an Acquity UPLC system (Waters, USA) equipped with an Eksigent ChromXP C18 column (3 μm, φ250 nm × 75 μm). The mobile phase consisted of solvent A (0.1% formic acid in acetonitrile, *v*/*v*) and solvent B (0.1% formic acid in water, *v*/*v*), and the elution conditions were as follows: 0.0–45.0 min, 95–70% A; 45.0–52.0 min, 70–20% A; 52.0–53.0 min, 20–95% A; 53.0–60.0 min, 95% A. The flow rate was set at 300 nL/min. The injection volume was 4 μL.

A Q-Exactive Orbitrap mass spectrometry instrument (Thermo Fisher Scientific, USA) was employed for the identification and quantification of ACEIPs from *U. intestinalis*. The conditions were set as follows: full MS, resolution 70,000, AGC 1e6, scan range 350–1600 *m*/*z*; dd-MS2, resolution 17,500, AGC 2e5, isolation window 2.0 *m*/*z*. The amino acid sequences of the peptides were determined by de novo sequencing using PEAKS Studio 6. Using a solid-phase method, synthetic peptides (purity >95% by HPLC) were obtained from China Peptides Co., Ltd (Shanghai, China).

3.9. In Vitro Digestion

In vitro digestion was performed according to the method of Cing et.al [50]. The purified peptide sample was mixed with 0.1 M phosphate buffer (pH 2.0) and pepsin (E/S 4%, *w*/*w*); the reaction was allowed to proceed for 2 h at 37 °C and stopped by heating at 100 °C for 10 min. Subsequently, the remaining suspension was adjusted to pH 8.0 with 5 M NaOH solution and further digested with trypsin (E/S 4%, *w*/*w*); the solution was incubated at 37 °C for 2 h and then heated at 100 °C for 10 min. The protein hydrolysate was centrifuged at 8000 rpm/min for 15 min, and the supernatant was assayed for ACE inhibitory activity.

3.10. Molecular Docking

The structures of the peptides were constructed with minimal energy using Chem Office 15.1 software (CambridgeSoft Co., USA). The crystal structure of human ACE (PDB: 1O8A) was obtained from RCSB PDB (Protein Data Bank; http://www.rcsb.org/pdb/home/home.do). Water molecules and lisinopril were removed from the ACE model before docking, whereas the cofactor zinc and chloride atoms were retained in the ACE model. The flexible docking tool of AutoDock 4.2 software (TSRI, USA) was used for docking. The docking runs were carried out as follows: coordinates x 41.586, y 37.383, and z 43.445; grid box size of 90 × 90 × 90 Å. The best molecular docking was considered the output based on docking scores and the binding-energy value derived from the best poses of the peptides interacting with ACE.

3.11. Statistical Analyses

Statistical analyses were performed in triplicate for each sample, and the results are expressed as the mean ± standard deviation (SD). One-way analysis of variance (ANOVA) was used to determine differences between the mean values of triplicate groups. Statistical significance was established at $P < 0.05$ with the least significant difference (LSD) procedure of SPSS version 19.0.

4. Conclusions

In this study, *U. intestinalis* proteins were enzymatic hydrolyzed using five different proteases, with the trypsin-digested hydrolysates exhibiting the highest ACE inhibition rate. Based on single-factor analysis and the RSM method, the optimum conditions were as follows: pH 8.4, temperature 28.5 °C, E/S 4.0%, substrate concentration 15 mg/mL, enzymolysis time 5.0 h. After a series of chromatographic separation and purification steps, two novel ACE inhibitory peptides were identified: Phe-Gly-Met-Pro-Leu-Asp-Arg (FGMPLDR; MW 834.41 Da) and Met-Glu-Leu-Val-Leu-Arg (MELVLR; MW 759.43 Da). The purified peptides displayed potent ACE inhibitory activity, with IC_{50} values of 219.35 and 236.85 μM, respectively. Based on in vitro digestion results, FGMPLDR and MELVLR demonstrated good stability for pepsin and trypsin digestion. Furthermore, we investigated interactions between the peptides and ACE by molecular docking, and the results indicated that hydrogen bonds and interaction with the Zn^{2+} of ACE contribute to the ACE inhibition activities of the two peptides. Overall, these peptides derived from trypsin hydrolysates of *U. intestinalis* may be considered for use in the industrial production of functional foods. However, further research such as antihypertensive activity experiments in mice should be performed.

Author Contributions: N.X. and X.C. conceived and designed the experiments; S.S. and X.X. performed the experiments; X.X., X.Z. and X.S. conducted the data analysis; S.S., X.Z. and N.X. wrote and revised the manuscript.

Funding: This work was supported by the Key Program of Natural Science Foundation of Zhejiang Province, China (LZ17D060001) and the Ningbo International Cooperation Program (2017D10019), and the program of Ningbo Science and Technology Bureau (2016C10034). This research was also sponsored by the KC Wong Magna Fund at Ningbo University.

Conflicts of Interest: The authors declare no conflict of interest.

References

1. Wang, J.; Hu, J.; Cui, J.; Bai, X.; Du, Y.; Miyaguchi, Y.; Lin, B. Purification and identification of a ACE inhibitory peptide from oyster proteins hydrolysate and the antihypertensive effect of hydrolysate in spontaneously hypertensive rats. *Food Chem.* **2008**, *111*, 302–308. [CrossRef] [PubMed]
2. Ngo, D.-H.; Kang, K.-H.; Ryu, B.; Vo, T.-S.; Jung, W.-K.; Byun, H.-G.; Kim, S.-K. Angiotensin-I converting enzyme inhibitory peptides from antihypertensive skate (*Okamejei kenojei*) skin gelatin hydrolysate in spontaneously hypertensive rats. *Food Chem.* **2015**, *174*, 37–43. [CrossRef] [PubMed]
3. Rai, A.K.; Sanjukta, S.; Jeyaram, K. Production of angiotensin I converting enzyme inhibitory (ACE-I) peptides during milk fermentation and their role in reducing hypertension. *Crit. Rev. Food Sci. Nutr.* **2017**, *57*, 2789–2800. [CrossRef] [PubMed]
4. Lahogue, V.; Réhel, K.; Taupin, L.; Haras, D.; Allaume, P. A HPLC-UV method for the determination of angiotensin I-converting enzyme (ACE) inhibitory activity. *Food Chem.* **2010**, *118*, 870–875. [CrossRef]
5. Chevillard, C.; Brown, N.L.; Mathieu, M.-N.; Laliberté, F.; Worcel, M. Differential effects of oral trandolapril and enalapril on rat tissue angiotensin-converting enzyme. *Eur. J. Pharmacol.* **1988**, *147*, 23–28. [CrossRef]
6. FitzGerald, R.J.; Murray, B.A.; Walsh, D.J. Hypotensive Peptides from Milk Proteins. *J. Nutr.* **2004**, *134*, 980S–988S. [CrossRef]
7. López-Fandiño, R.; Otte, J.; van Camp, J. Physiological, chemical and technological aspects of milk-protein-derived peptides with antihypertensive and ACE-inhibitory activity. *Int. Dairy J.* **2006**, *16*, 1277–1293. [CrossRef]
8. Fu, Y.; Young, J.F.; Løkke, M.M.; Lametsch, R.; Aluko, R.E.; Therkildsen, M. Revalorisation of bovine collagen as a potential precursor of angiotensin I-converting enzyme (ACE) inhibitory peptides based on in silico and in vitro protein digestions. *J. Funct. Foods* **2016**, *24*, 196–206. [CrossRef]
9. Lau, C.C.; Abdullah, N.; Shuib, A.S.; Aminudin, N. Novel angiotensin I-converting enzyme inhibitory peptides derived from edible mushroom *Agaricus bisporus* (J.E. Lange) Imbach identified by LC–MS/MS. *Food Chem.* **2014**, *148*, 396–401. [CrossRef] [PubMed]
10. Chen, J.; Liu, S.; Ye, R.; Cai, G.; Ji, B.; Wu, Y. Angiotensin-I converting enzyme (ACE) inhibitory tripeptides from rice protein hydrolysate: Purification and characterization. *J. Funct. Foods* **2013**, *5*, 1684–1692. [CrossRef]
11. Tsai, J.-S.; Chen, J.-L.; Pan, B.S. ACE-inhibitory peptides identified from the muscle protein hydrolysate of hard clam (*Meretrix lusoria*). *Process Biochem.* **2008**, *43*, 743–747. [CrossRef]
12. Cao, D.; Lv, X.; Xu, X.; Yu, H.; Sun, X.; Xu, N. Purification and identification of a novel ACE inhibitory peptide from marine alga *Gracilariopsis lemaneiformis* protein hydrolysate. *Eur. Food Res. Technol.* **2017**, *243*, 1829–1837. [CrossRef]
13. Larsen, R.; Eilertsen, K.-E.; Elvevoll, E.O. Health benefits of marine foods and ingredients. *Biotechnol. Adv.* **2011**, *29*, 508–518. [CrossRef] [PubMed]
14. Fitzgerald, C.; Gallagher, E.; Tasdemir, D.; Hayes, M. Heart Health Peptides from Macroalgae and Their Potential Use in Functional Foods. *J. Agric. Food Chem.* **2011**, *59*, 6829–6836. [CrossRef] [PubMed]
15. Suetsuna, K.; Maekawa, K.; Chen, J.-R. Antihypertensive effects of *Undaria pinnatifida* (wakame) peptide on blood pressure in spontaneously hypertensive rats. *J. Nutr. Biochem.* **2004**, *15*, 267–272. [CrossRef]
16. Suetsuna, K. Purification and identification of angiotensin I-converting enzyme inhibitors from the red alga Porphyra yezoensis. *J. Mar. Biotechnol.* **1998**, *6*, 163–167.
17. Paiva, L.; Lima, E.; Neto, A.I.; Baptista, J. Isolation and characterization of angiotensin I-converting enzyme (ACE) inhibitory peptides from Ulva rigida C. Agardh protein hydrolysate. *J. Funct. Foods* **2016**, *26*, 65–76. [CrossRef]
18. Lu, J.; Ren, D.-F.; Xue, Y.-L.; Sawano, Y.; Miyakawa, T.; Tanokura, M. Isolation of an Antihypertensive Peptide from Alcalase Digest of Spirulina platensis. *J. Agric. Food Chem.* **2010**, *58*, 7166–7171. [CrossRef] [PubMed]
19. Ibrahim, D.; Lim, S.-H. In vitro antimicrobial activities of methanolic extract from marine alga Enteromorpha intestinalis. *Asian Pac. J. Trop. Biomed.* **2015**, *5*, 785–788. [CrossRef]
20. Martins, I.; Marques, J.C. A Model for the Growth of Opportunistic Macroalgae (*Enteromorpha* sp.) in Tidal Estuaries. *Estuar. Coast. Shelf Sci.* **2002**, *55*, 247–257. [CrossRef]
21. Bäck, S.; Lehvo, A.; Blomster, J. Mass occurrence of unattached Enteromorpha intestinalis on the Finnish Baltic Sea Coast. *Ann. Bot. Fenn.* **2000**, *37*, 155–161.

22. Alström-Rapaport, C.; Leskinen, E.; Pamilo, P. Seasonal variation in the mode of reproduction of *Ulva intestinalis* in a brackish water environment. *Aquat. Bot.* **2010**, *93*, 244–249. [CrossRef]
23. Peasura, N.; Laohakunjit, N.; Kerdchoechuen, O.; Wanlapa, S. Characteristics and antioxidant of *Ulva intestinalis* sulphated polysaccharides extracted with different solvents. *Int. J. Biol. Macromol.* **2015**, *81*, 912–919. [CrossRef] [PubMed]
24. Kim, D.-H.; Lee, S.-B.; Jeong, G.-T. Production of reducing sugar from *Enteromorpha intestinalis* by hydrothermal and enzymatic hydrolysis. *Bioresour. Technol.* **2014**, *161*, 348–353. [CrossRef] [PubMed]
25. Sheih, I.C.; Fang, T.J.; Wu, T.-K. Isolation and characterisation of a novel angiotensin I-converting enzyme (ACE) inhibitory peptide from the algae protein waste. *Food Chem.* **2009**, *115*, 279–284. [CrossRef]
26. Kristinsson, H.G.; Rasco, B.A. Fish Protein Hydrolysates: Production, Biochemical, and Functional Properties. *Crit. Rev. Food Sci. Nutr.* **2000**, *40*, 43–81. [CrossRef] [PubMed]
27. Lee, J.K.; Jeon, J.-K.; Byun, H.-G. Antihypertensive effect of novel angiotensin I converting enzyme inhibitory peptide from chum salmon (*Oncorhynchus keta*) skin in spontaneously hypertensive rats. *J. Funct. Foods* **2014**, *7*, 381–389. [CrossRef]
28. Derrien, M.; Badr, A.; Gosselin, A.; Desjardins, Y.; Angers, P. Optimization of a green process for the extraction of lutein and chlorophyll from spinach by-products using response surface methodology (RSM). *LWT Food Sci. Technol.* **2017**, *79*, 170–177. [CrossRef]
29. Ma, M.-S.; Bae, I.Y.; Lee, H.G.; Yang, C.-B. Purification and identification of angiotensin I-converting enzyme inhibitory peptide from buckwheat (*Fagopyrum esculentum* Moench). *Food Chem.* **2006**, *96*, 36–42. [CrossRef]
30. Rho, S.J.; Lee, J.-S.; Chung, Y.I.; Kim, Y.-W.; Lee, H.G. Purification and identification of an angiotensin I-converting enzyme inhibitory peptide from fermented soybean extract. *Process Biochem.* **2009**, *44*, 490–493. [CrossRef]
31. Asoodeh, A.; Homayouni-Tabrizi, M.; Shabestarian, H.; Emtenani, S.; Emtenani, S. Biochemical characterization of a novel antioxidant and angiotensin I-converting enzyme inhibitory peptide from *Struthio camelus* egg white protein hydrolysis. *J. Food Drug Anal.* **2016**, *24*, 332–342. [CrossRef]
32. Jimsheena, V.K.; Gowda, L.R. Colorimetric, High-Throughput Assay for Screening Angiotensin I-Converting Enzyme Inhibitors. *Anal. Chem.* **2009**, *81*, 9388–9394. [CrossRef]
33. Ryan, J.T.; Ross, R.P.; Bolton, D.; Fitzgerald, G.F.; Stanton, C. Bioactive Peptides from Muscle Sources: Meat and Fish. *Nutrients* **2011**, *3*, 765–791. [CrossRef]
34. Li, M.; Xia, S.; Zhang, Y.; Li, X. Optimization of ACE inhibitory peptides from black soybean by microwave-assisted enzymatic method and study on its stability. *LWT* **2018**, *98*, 358–365. [CrossRef]
35. Lee, J.K.; Jeon, J.-K.; Byun, H.-G. Effect of angiotensin I converting enzyme inhibitory peptide purified from skate skin hydrolysate. *Food Chem.* **2011**, *125*, 495–499. [CrossRef]
36. Ruiz, J.Á.G.; Ramos, M.; Recio, I. Angiotensin converting enzyme-inhibitory activity of peptides isolated from Manchego cheese. Stability under simulated gastrointestinal digestion. *Int. Dairy J.* **2004**, *14*, 1075–1080. [CrossRef]
37. Haque, E.; Chand, R. Antihypertensive and antimicrobial bioactive peptides from milk proteins. *Eur. Food Res. Technol.* **2008**, *227*, 7–15. [CrossRef]
38. Yu, F.; Zhang, Z.; Luo, L.; Zhu, J.; Huang, F.; Yang, Z.; Tang, Y.; Ding, G. Identification and Molecular Docking Study of a Novel Angiotensin-I Converting Enzyme Inhibitory Peptide Derived from Enzymatic Hydrolysates of *Cyclina sinensis*. *Mar. Drugs* **2018**, *16*, 411. [CrossRef]
39. Wu, J.; Ding, X. Characterization of inhibition and stability of soy-protein-derived angiotensin I-converting enzyme inhibitory peptides. *Food Res. Int.* **2002**, *35*, 367–375. [CrossRef]
40. Ngo, D.-H.; Vo, T.-S.; Ryu, B.; Kim, S.-K. Angiotensin- I- converting enzyme (ACE) inhibitory peptides from Pacific cod skin gelatin using ultrafiltration membranes. *Process Biochem.* **2016**, *51*, 1622–1628. [CrossRef]
41. Mirzaei, M.; Mirdamadi, S.; Ehsani, M.R.; Aminlari, M. Production of antioxidant and ACE-inhibitory peptides from *Kluyveromyces marxianus* protein hydrolysates: Purification and molecular docking. *J. Food Drug Anal.* **2018**, *26*, 696–705. [CrossRef]
42. Chaudhary, S.; Vats, I.D.; Chopra, M.; Biswas, P.; Pasha, S. Effect of varying chain length between P1 and P1′ position of tripeptidomimics on activity of angiotensin-converting enzyme inhibitors. *Bioorg. Med. Chem. Lett.* **2009**, *19*, 4364–4366. [CrossRef]

43. Wu, Q.; Jia, J.; Yan, H.; Du, J.; Gui, Z. A novel angiotensin-I converting enzyme (ACE) inhibitory peptide from gastrointestinal protease hydrolysate of silkworm pupa (*Bombyx mori*) protein: Biochemical characterization and molecular docking study. *Peptides* **2015**, *68*, 17–24. [CrossRef]
44. Abdelhedi, O.; Nasri, R.; Jridi, M.; Mora, L.; Oseguera-Toledo, M.E.; Aristoy, M.-C.; Amara, I.B.; Toldrá, F.; Nasri, M. In silico analysis and antihypertensive effect of ACE-inhibitory peptides from smooth-hound viscera protein hydrolysate: Enzyme-peptide interaction study using molecular docking simulation. *Process Biochem.* **2017**, *58*, 145–159. [CrossRef]
45. Rawendra, R.D.; Aisha; Chang, C.-I.; Aulanni'am; Chen, H.-H.; Huang, T.-C.; Hsu, J.-L. A novel angiotensin converting enzyme inhibitory peptide derived from proteolytic digest of Chinese soft-shelled turtle egg white proteins. *J. Proteom.* **2013**, *94*, 359–369. [CrossRef]
46. Pan, D.; Guo, H.; Zhao, B.; Cao, J. The molecular mechanisms of interactions between bioactive peptides and angiotensin-converting enzyme. *Bioorg. Med. Chem. Lett.* **2011**, *21*, 3898–3904. [CrossRef]
47. Wu, H.; Xu, N.; Sun, X.; Yu, H.; Zhou, C. Hydrolysis and purification of ACE inhibitory peptides from the marine microalga *Isochrysis galbana*. *J. Appl. Phycol.* **2015**, *27*, 351–361. [CrossRef]
48. Shalaby, S.; Zakora, M.; Otte, J. Performance of two commonly used angiotensin-converting enzyme inhibition assays using FA-PGG and HHL as substrates. *J. Dairy Res.* **2006**, *73*, 178–186. [CrossRef]
49. Henda, Y.B.; Labidi, A.; Arnaudin, I.; Bridiau, N.; Delatouche, R.; Maugard, T.; Piot, J.-M.; Sannier, F.; Thiéry, V.; Bordenave-Juchereau, S. Measuring Angiotensin-I Converting Enzyme Inhibitory Activity by Micro Plate Assays: Comparison Using Marine Cryptides and Tentative Threshold Determinations with Captopril and Losartan. *J. Agric. Food Chem.* **2013**, *61*, 10685–10690. [CrossRef]
50. Cinq-Mars, C.D.; Hu, C.; Kitts, D.D.; Li-Chan, E.C.Y. Investigations into Inhibitor Type and Mode, Simulated Gastrointestinal Digestion, and Cell Transport of the Angiotensin I-Converting Enzyme–Inhibitory Peptides in Pacific Hake (*Merluccius productus*) Fillet Hydrolysate. *J. Agric. Food Chem.* **2008**, *56*, 410–419. [CrossRef]

Article

Partial Characterization of Protease Inhibitors of *Ulva ohnoi* and Their Effect on Digestive Proteases of Marine Fish

Antonio Jesús Vizcaíno *, Alba Galafat, María Isabel Sáez, Tomás Francisco Martínez and Francisco Javier Alarcón

Departamento de Biología y Geología, Escuela Superior de Ingeniería, Ceimar-Universidad de Almería. La Cañada de San Urbano, 04120 Almería, Spain; albagalafat@gmail.com (A.G.); msc880@ual.es (M.I.S.); tomas@ual.es (T.F.M.); falarcon@ual.es (F.J.A.)

* Correspondence: avt552@ual.es; Tel.: +34-950-015954

Received: 30 April 2020; Accepted: 16 June 2020; Published: 18 June 2020

Abstract: This piece of research evaluates the presence of protease inhibitors in the macroalga *Ulva ohnoi* and provides an initial overview of their mode of action. The ability of *Ulva* protease inhibitors to inhibit digestive proteases of three marine fish species, as well as their capacity to hamper the hydrolysis of a reference protein by those fish proteases, were assessed. In addition, thermal stability and the mode of inhibition on trypsin and chymotrypsin were also studied. Dose-response inhibition curves and in vitro protein hydrolysis assays revealed a noticeable inhibition of fish enzymes when *Ulva* concentration increased in the assay. The thermal treatment of *Ulva* reduced markedly the inhibitory effect on fish digestive protease. Finally, Lineweaver–Burk plots indicated that trypsin and chymotrypsin inhibition consisted of a mixed-type inhibition mechanism in which the inhibitory effect depends on *Ulva* concentration. Overall, the results confirmed the presence of protease inhibitors in *Ulva*, though heat treatment was enough for inactivating these compounds.

Keywords: inhibitor; macroalgae; marine fish; protease; *Ulva ohnoi*

---

1. Introduction

Anti-nutritional factors (ANFs) can be defined as substances that, by themselves or through their metabolic products, can exert negative effects on food utilization and interfere with the normal growth, reproduction and health of fish [1]. From a nutritional point of view, the presence of these compounds in diets is responsible for the deleterious effects on the absorption of nutrients and micronutrients, which may interfere with the normal functioning of certain organs [2]. This fact is one of the most important issues derived from using novel dietary ingredients in aquaculture, as well as one of the main drawbacks that limits their practical application in formulated feeds [3,4].

In general, ANFs have been mainly related to plant-derived feedstuffs, and they comprise a wide variety of compounds, such as protease inhibitors, phytohemagglutinin, lectins, phytic acid, saponins, phytoestrogens or antivitamins [1,5]. Although less known, recent studies have also documented the presence of these substances in seaweeds, considered currently a potential alternative ingredient for farmed fish [6,7].

Seaweeds have drawn the attention of researchers not only as an important source of dietary protein, but also as functional ingredients in aquafeeds [8]. More specially, some species of the genus *Ulva* have been successfully evaluated as a dietary ingredient in some farmed fish species, such as gilthead seabream (*Sparus aurata*) [9,10], Senegalese sole (*Solea senegalensis*) [8,11] or seabass (*Dicentrarchus labrax*) [12], with promising results in terms of growth, survival and nutrient utilization. Despite the health benefits reported for seaweeds, some studies have described that the dietary inclusion level of algae above 20% yields detrimental effects on fish growth and other zootechnical parameters. It has been suggested that these effects could be attributed to the existence of ANFs, which might affect the bioavailability and/or digestibility of nutrients [13].

In this regard, Oliveira et al. [14] and Maehre, [6] confirmed the presence of lectins, trypsin and alpha-amylase inhibitors, as well as phytic acid, in some species of marine algae, although polyphenolic compounds are considered the substances most frequently linked to their antinutritional effects [6,14]. It has been described that the presence of lectins in feed alters the intestinal epithelium, resulting in the over-secretion of mucus that may impair the enzymatic and absorptive capacity of fish [15], altogether leading to reduced growth [5]. Phytic acid and polyphenols bind to proteins and polysaccharides producing insoluble high-molecular complexes, a fact that reduces nutrient bioavailability and consequent nutrient deficiency, such as that described for methionine, which is essential for lipid metabolism [15,16]. In addition, other antinutritional compounds of seaweeds, such as phytates, can inhibit the action of gastrointestinal enzymes like tyrosinase, trypsin, pepsin, lipase and amylase [2].

It is also worth mentioning the existence of protease inhibitors, substances that bind to proteolytic enzymes causing not only reduced proteolysis, but also increased pancreatic secretion as an attempt to overcome these antinutritional effects [17]. Despite the evidence suggesting the presence of protease inhibitors in seaweeds and their possible effects on the digestive physiology of aquacultured fish [7,13,17], scarce research is available regarding the characterization of their effects. In this context, this this research aims to assess the presence of protease inhibitors in *Ulva ohnoi*, evaluating the effects of such inhibitors on fish digestive proteases, characterizing their mode of inhibition, and exploring potential strategies to reduce their detrimental effects on fish digestive enzymes.

2. Results

2.1. Inhibition Assay

The inhibitory effect of *Ulva ohnoi* on the digestive proteases of gilthead seabream, Senegalese sole and seabass is shown in Figure 1. Dose-response inhibition curves were obtained by measuring the reduction in proteolytic activity on a standardized fish intestinal extract following incubation with different volumes of *Ulva* extract. The results confirmed the presence of protease inhibitor in crude *Ulva* (CR-*Ulva*) able to inhibit up to 70% of Senegalese sole's proteolytic activity, and by 65% of protease activity in the other two fish species. It was also found that the amount of *Ulva* able to cause the inhibition of 50% digestive protease activity (IC_{50}) ranged from 0.6 to 0.9 mg *Ulva* per unit of proteolytic activity (UA). On the other hand, autoclaved *Ulva* (heat treated; HT-*Ulva*) reduced significantly its inhibitory capacity (less than 20% inhibition in all cases) compared to CR-*Ulva* ($p < 0.05$).

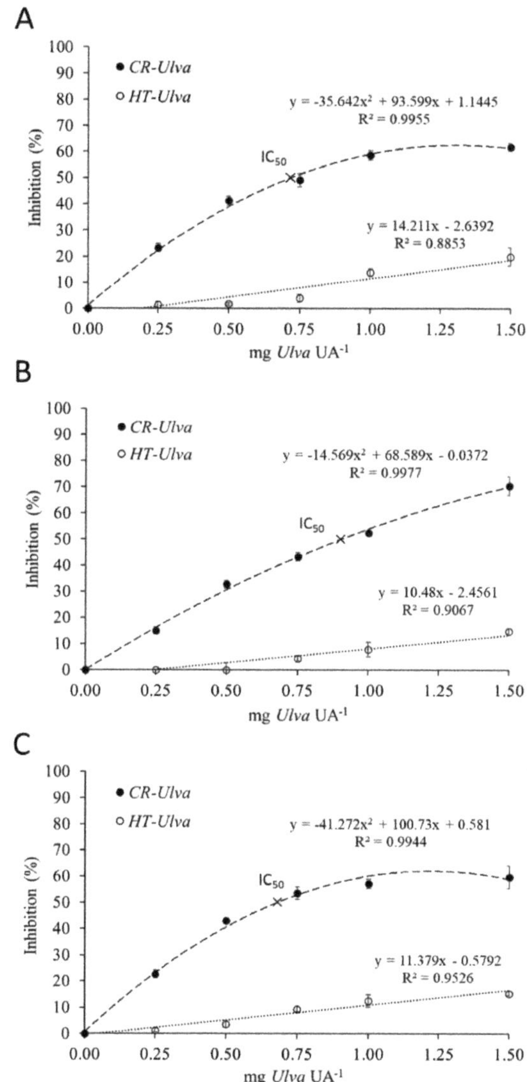

Figure 1. Dose-response inhibition plot of *S. aurata* (**A**), *S. senegalensis* (**B**) and *D. labrax* (**C**) intestinal proteases activity following the incubation with increasing concentrations of crude (CR-*Ulva*) and heat-treated *Ulva* extracts (HT-*Ulva*). Each point represents the mean of triplicates ± SD. IC_{50}: mg of *Ulva* needed to cause 50% protease inhibition.

Zymograms of fish digestive proteases after incubation with *Ulva* extract are shown in Figure 2. The effects of protease inhibitors were compared to a control without *Ulva* (lane 1). Noticeable reduction in the intensity of the active fractions (with proteolytic activity) was evidenced after incubation with *Ulva* (lanes 2 to 4). Differences in the inhibitory effect were observed among the three fish species. In the case of gilthead seabream, a progressive decrease in the intensity of all active proteases was observed as the *Ulva* concentration increased, whereas protease inhibition in the other fish was selective for some specific active fractions.

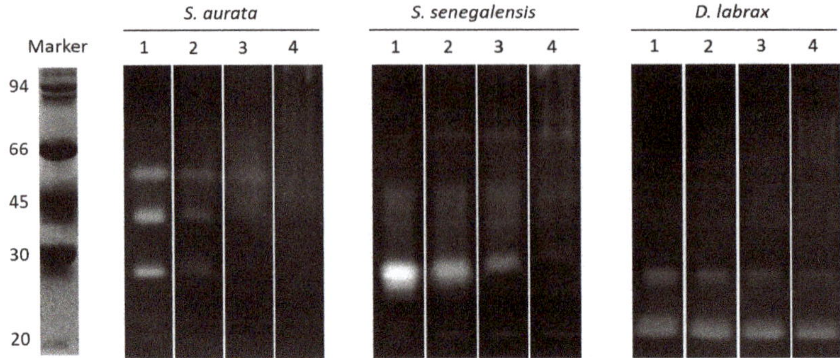

Figure 2. Substrate-SDS-PAGE (sodium dodecyl sulphate polyacrylamide gel electrophoresis) obtained after incubation of *S. aurata*, *S. senegalensis* and *D. labrax* intestinal extracts with different concentrations of *Ulva* extract. Lane 1: control without inhibitor (distilled water was used instead of *Ulva* extract); lane 2: 500 µg *Ulva* per unit of proteolytic activity (UA); lane 3: 1000 µg *Ulva* per UA; lane 4: 1500 µg *Ulva* per UA. The molecular mass markers (kDa) were phosphorylase b (94), bovine serum albumin (66), ovalbumin (45), carbonic anhydrase (30), and soybean trypsin inhibitor (20).

2.2. *In Vitro Casein Hydrolysis*

Proteinograms of casein hydrolyzed by fish digestive enzymes are shown in Figure 3. All the main casein fractions (34, 26 and 23 kDa), corresponding to α, β and κ subunits, were hydrolyzed after 30 min.

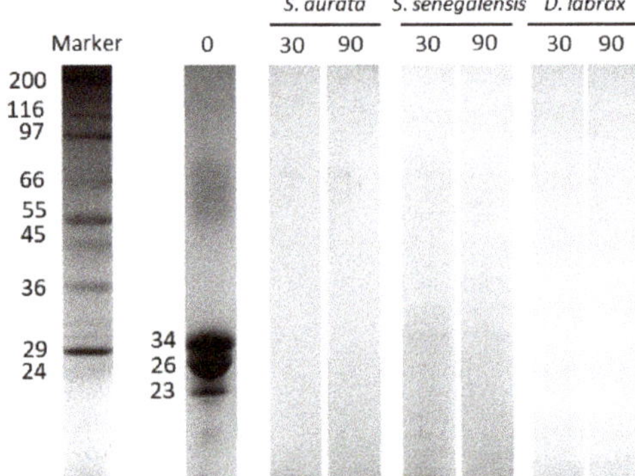

Figure 3. The time-course of in vitro proteolysis of casein by *S. aurata*, *S. senegalensis* and *D. labrax* digestive proteases in the absence of any inhibitory extract. Images show SDS-PAGE hydrolysis patterns obtained at different sampling times (0, 30, and 90 min). Numbers on the left of the electrophoresis gel stand for the molecular mass of proteins fractions (kDa).

Figure 4 shows the time-course protein hydrolysis of casein by the fish digestive proteases in the presence of *Ulva* extracts. Overall, CR-*Ulva* hampered the capacity of fish proteases to hydrolyze casein. The fate of the casein fractions throughout the in vitro assay was partial and less marked compared to the same assay carried out without *Ulva* (Figure 3). It was also observed that the inhibitory effect was dose-dependent, given that the higher CR-*Ulva* concentration in the in vitro assay, the lower the hydrolytic capacity of fish digestive enzymes against casein. The inhibitory effect of *Ulva* extracts on digestive proteases was reduced remarkably owing to thermal treatment, as revealed by the proteinograms shown in Figure 4 (lanes at the right). Thus, heat-treated seaweed extracts (HT-*Ulva*) were affected casein hydrolysis to a lesser extent than untreated *Ulva* (CR-*Ulva*), in such a way that none of the casein subunits remained after 90 min in any of the assays carried out with HT-*Ulva*. However, intraspecific and dose-dependent differences were found, and thus increasing concentrations of HT-*Ulva* also reduced casein hydrolysis, not least for seabream and seabass digestive proteases.

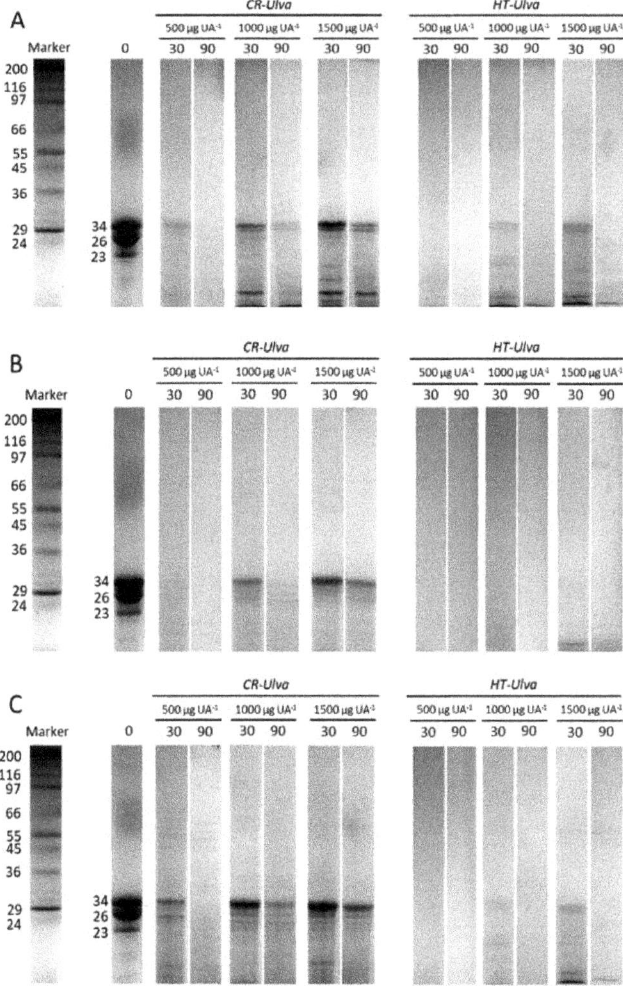

Figure 4. The time-course of in vitro proteolysis of casein by *S. aurata* (**A**), *S. senegalensis* (**B**) and *D. labrax* (**C**) digestive proteases in the presence of increasing concentrations of crude (CR-*Ulva*) and heat-treated (HT-*Ulva*) *Ulva*. The images show SDS-PAGE hydrolysis patterns obtained at different sampling times (0, 30, and 90 min). The numbers on the left of the electrophoresis gels stand for the molecular mass of the protein fractions (kDa).

The estimated values of the coefficient of protein degradation (CPD) and total amino acids released after the in vitro hydrolysis of casein are given in Table 1. The CR-*Ulva* lowered the rate of protein degradation compared to the assay performed without *Ulva* ($p < 0.05$). The lowest CPD values were observed for the highest concentrations of CR-*Ulva* in the reaction mixture (1500 µg per UA). In parallel, the amount of free amino acids released was also affected by the interaction of *Ulva* extracts with fish proteases, especially for highest concentrations of CR-Ulva. The impact of HT-*Ulva* on *S. aurata* and

S. senegalensis digestive protease activity was negligible, since neither the CPD nor free amino acids that were released were different from the controls. However, a certain residual inhibitory effect of HT-*Ulva* was observed on *D. labrax* enzyme extracts.

Table 1. The values of the estimated coefficient of protein degradation (CPD) and total amino acids released measured after 90 min of in vitro hydrolysis of casein by fish digestive proteases in the presence of different concentrations (0, 500, 1000 and 1500 µg *Ulva* per UA) of crude (CR-*Ulva*) and heat-treated (HT-*Ulva*) *Ulva ohnoi*.

Fish vs. *Ulva* Concentration	CPD (%)	Total Amino Acids Released (g 100 g protein^{-1})
Sparus aurata		
Control	91.6 ± 2.3 ^{d}	22.1 ± 2.3 ^{d}
CR-*Ulva* 500	77.9 ± 2.0 ^{c}	15.4 ± 0.1 ^{c}
CR-*Ulva* 1000	62.6 ± 4.1 ^{b}	11.1 ± 0.7 ^{b}
CR-*Ulva* 1500	45.9 ± 2.0 ^{a}	7.4 ± 2.4 ^{a}
HT-*Ulva* 500	91.4 ± 2.1 ^{d}	20.7 ± 1.0 ^{d}
HT-*Ulva* 1000	90.6 ± 1.0 ^{d}	20.8 ± 1.6 ^{d}
HT-*Ulva* 1500	89.2 ± 1.7 ^{d}	20.7 ± 2.7 ^{d}
p	<0.001	<0.001
Solea senegalensis		
Control	91.9 ± 2.2 ^{d}	22.1 ± 1.5 ^{c}
CR-*Ulva* 500	84.3 ± 0.4 ^{c}	19.6 ± 0.8 ^{c}
CR-*Ulva* 1000	66.1 ± 1.5 ^{b}	12.8 ± 0.3 ^{b}
CR-*Ulva* 1500	41.8 ± 4.4 ^{a}	8.5 ± 1.0 ^{a}
HT-*Ulva* 500	89.9 ± 1.2 ^{d}	21.9 ± 1.8 ^{c}
HT-*Ulva* 1000	89.1 ± 2.1 ^{d}	21.5 ± 2.0 ^{c}
HT-*Ulva* 1500	88.2 ± 1.3 ^{d}	20.8 ± 0.8 ^{c}
p	<0.001	<0.001
Dicentrarchus labrax		
Control	96.2 ± 1.2 ^{c}	22.3 ± 2.5 ^{d}
CR-*Ulva* 500	91.2 ± 0.8 ^{b}	16.3 ± 1.0 ^{c}
CR-*Ulva* 1000	69.9 ± 1.4 ^{a}	11.8 ± 0.3 ^{b}
CR-*Ulva* 1500	66.6 ± 2.5 ^{a}	8.2 ± 0.3 ^{a}
HT-*Ulva* 500	95.1 ± 1.8 ^{b,c}	22.0 ± 2.3 ^{d}
HT-*Ulva* 1000	94.1 ± 3.2 ^{b,c}	21.3 ± 0.6 ^{d}
HT-*Ulva* 1500	93.4 ± 1.4 ^{b}	20.1 ± 0.6 ^{d}
p	<0.001	<0.001

The values are mean ± SD of triplicate determinations. Within each fish species, the values in the same column with different lowercase letters indicate significant differences ($p < 0.05$) owing to *Ulva* extract. ^{a,b,c,d} the superscripts indicate the treatments with significant differences.

2.3. Thermal Stability of Protease Inhibitors

The influence of a thermal treatment on the inhibitory capacity of *Ulva* on trypsin activity was evaluated. A noticeable reduction in the inhibitory effects of *Ulva* was observed as the temperature increased (Figure 5). Protease inhibitors of the *Ulva* extracts were stable at room temperature, but their inhibitory effect significantly reduced ($p < 0.05$) above 40 °C. The exposure of *Ulva* extract at 80 °C for 30 min reduced their inhibitory activity up to 50% compared to the untreated controls.

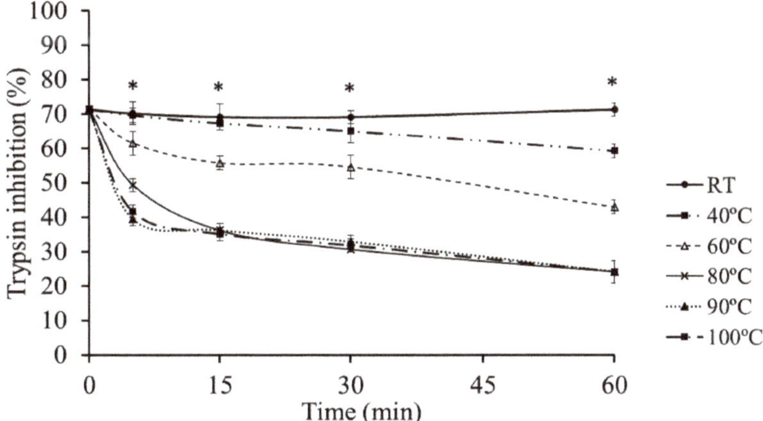

Figure 5. The effects of thermal treatments on the inhibitory effect of *Ulva* on trypsin activity. * Denote significant differences among treatments ($p < 0.05$).

The existence of proteinaceous complexes consisting of *Ulva* protease inhibitors and trypsin is revealed in Figure 6. The commercial trypsin showed three protein fractions (24, 24.4, and 26 kDa, lane 1), whereas the protein pattern of the *Ulva* extract contained several protein fractions (lane 2). The incubation of trypsin with crude *Ulva* before SDS-PAGE (sodium dodecyl sulphate polyacrylamide gel electrophoresis) yielded a characteristic protein profile in which the 24 and 26 kDa fractions of trypsin disappeared. Instead, a new 31 kDa fraction appeared (lanes 3 to 5), which was not present either in trypsin or in *Ulva*. However, this 31 kDa protein fraction did not appear when trypsin was incubated with heat-treated *Ulva* (lanes 6 to 8).

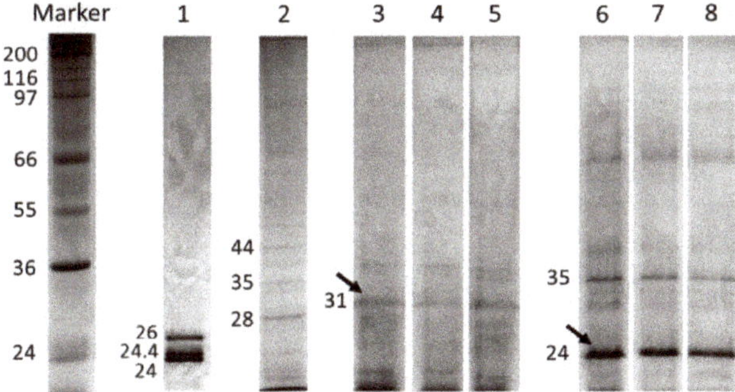

Figure 6. SDS-PAGE carried out to detect trypsin-*Ulva* inhibitor complexes. Lane 1 = trypsin (1 mg mL^{-1}). Lane 2 = *Ulva* (0.1 g mL^{-1}). Lanes 3 to 5 = trypsin preincubated with CR-*Ulva* during 0, 30 and 60 min. Lanes 6 to 8 = trypsin pre-incubated with heat-treated *Ulva* during 0, 30 and 60 min.

2.4. Kinetic Parameter's

The data obtained for the kinetic studies performed with trypsin and chymotrypsin are shown in Figure 7. The results demonstrated that *Ulva* protease inhibitors yielded a potential mixed type inhibition, as revealed by the decrease in V_{max} as K_m increased, compared to the apparent kinetic parameters obtained for both trypsin and chymotrypsin activities in the absence of *Ulva* (Table 2).

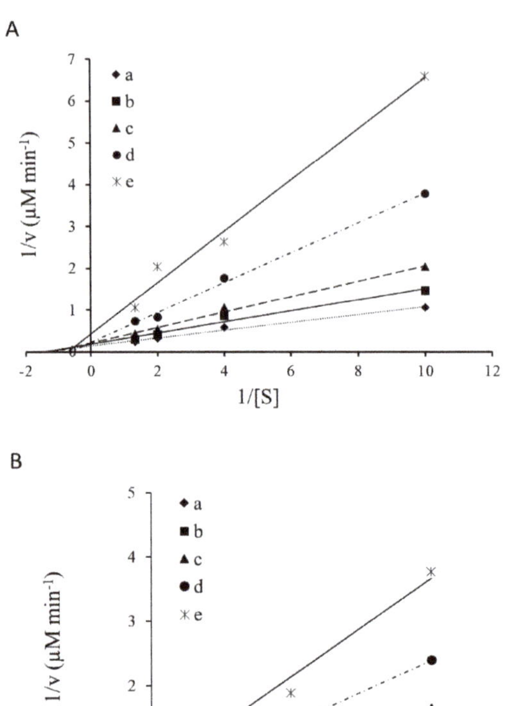

Figure 7. Lineweaver–Burk plots showing the inhibition of trypsin (**A**) and chymotrypsin (**B**) by *Ulva* protease inhibitors (a: 0; b: 200; c: 300; d: 400 and e: 500 μg *Ulva* per μg enzyme).

Table 2. Kinetic analysis of trypsin and chymotrypsin enzymes in the presence of *Ulva*.

***Ulva* Concentration**	**Trypsin**		**Chymotrypsin**	
	K_m (mM)	**V_{max} (μMmol pNA min^{-1})**	**K_m (mM)**	**V_{max} (μMmol pNA min^{-1})**
Without *Ulva*	0.65	7.03	0.15	5.92
200 μg *Ulva* per μg enzyme	0.73	5.54	0.20	5.57
300 μg *Ulva* per μg enzyme	0.87	4.71	0.21	3.15
400 μg *Ulva* per μg enzyme	1.58	4.43	0.23	2.38
500 μg *Ulva* per μg enzyme	2.53	4.12	0.25	1.65

K_m: Michaelis constant; V_{max}: maximum rate of reaction; pNA: para-nitroaniline.

3. Discussion

Seaweeds are considered a promising alternative ingredient for aquafeeds. In addition to having nutrients with potential quantitative interest, the presence of biologically active compounds like polysaccharides, pigments (chlorophylls and carotenoids), sterols, polyphenols, and vitamins also makes seaweeds a valuable functional ingredient for aquafeeds [18,19]. However, it has been reported that macroalgae contain several anti-nutritional factors as well, such as lectins and protease inhibitors that might interfere with digestive processes [8]. In this regard, the present study explores the presence of antinutritional factors in *Ulva ohnoi* with a potential inhibitory effect on fish digestive proteases and provides an initial overview of their mode of action on those proteolytic enzymes.

Related to *Ulva* ANFs, the results of the present study confirmed the existence of compounds able to reduce the digestive proteolytic activity of different marine fish species. This fact agrees with previous studies pointing to the existence of protease inhibitors in some macroalgae species, such as *Ulva rigida, Ulva ohnoi, Gracilaria cornea* and *Sargassum* sp. [7,8,17]. In our study, inhibition plots and zymograms illustrate the response of fish proteases after incubation with crude *Ulva ohnoi*. Seabream, Senegalese sole and seabass digestive proteases showed susceptibility to *Ulva* ANFs, although a high concentration of *Ulva* was needed to cause high inhibition (>50% of protease inhibition). According to the equations obtained from the inhibition assays, the amount of *Ulva* required to reach IC_{50} would represent a dietary inclusion of approximately 40–53% in feeds, which are unusual levels for feed formulation. From a practical point of view, for 40 g fish consuming feeds containing 15% *Ulva* at an intake rate of 3% of its body weight, a 20% reduction in their digestive protease activity would be expected. In general, fish have physiological mechanisms aimed at compensating the effects of dietary antinutrients [20], although the influence of these compounds is species-specific [21] and may depend on different factors like fish physiology, macroalgae species, duration of feeding period with seaweed-supplemented diets, and the dietary inclusion level [8]. In this regard, there are numerous studies reporting the utilization of dietary seaweeds without compromising growth performance [19,22]. However, others described some negative effects on fish growth, even using seaweed biomass at low dietary inclusion level [8].

The effect of *Ulva* inhibitors on fish digestive proteases were also evidenced in zymograms. Protease inhibition caused by CR-*Ulva* on gilthead seabream enzymes could be classified as "unspecific", owing to the fact that all the protease fractions visualized in the gels were affected similarly. However, the inhibitory effect seemed to be more "specific" in Senegalese sole and seabass, taking into account that the active fractions with molecular masses below 30 kDa were inhibited even with the lowest amount of crude *Ulva*, whereas heavier proteases were inhibited only using the highest concentration assayed. This reduction in protease activity may negatively affect feed intake and nutrient digestibility in fish [23]. In this context, the in vitro digestive simulations also confirmed that CR-*Ulva* hampered the hydrolysis of standard (casein) by digestive proteases of three different species of aquaculture fish. Thus, a clear reduction both in CPD values and the amount of amino acids released was evidenced when the concentration of CR-*Ulva* in the in vitro assay was increased. Considered together, both findings clearly indicate lower protein hydrolysis and also reflect a significant reduction in the hydrolytic action of both the digestive endo- and exo-proteases of fish [24].

The deactivation of ANFs is an important issue in raw materials processing [25]. Basically, ANFs can be divided into two groups: (i) heat-labile ANFs, including protease inhibitors, phytates and lectins; and (ii) heat-stable ANFs, represented by saponins, non-starch polysaccharides and some phenolic compounds [2]. Heat treatment is a simple procedure for inactivating ANFs and improving the nutritional value of raw protein feedstuffs [26]. The results in our study indicate that thermal treatment is effective when it comes to inactivating *Ulva ohnoi* ANFs that affect fish digestive proteases. The degree of ANFs inactivation depends on factors like temperature, time, particle size, and moisture conditions [2]. In fact, both time and temperature should be controlled carefully in order to minimize losses of nutritional value of a given feed ingredient (for instance, lower availability of amino acids and vitamins, and reduced protein bioaccessibility) as a result of excessive heat denaturation [15,27]. In this regard, the effect of the thermal treatment on the capacity of *Ulva* to inhibit trypsin activity indicates that such inhibitors are susceptible to relatively slight thermal treatment. A thermal treatment of 80 °C for 15 min reduced the inhibitory capacity by 50%, and above 75% as prolonged times were applied. In agreement, proteinograms (Figure 6, lanes 6 to 8) also confirmed that *Ulva* protease inhibitors are thermolabile, owing to the lack of detection of the 31 kDa proteinaceous complex following the thermal treatment of *Ulva*. These results suggest that temperatures reached during the standard industrial processing, for instance, in the extrusion of feeds, would be enough for minimizing the inhibitory capacity of *Ulva*. Both the preconditioning of the ingredient mixture and the friction forces of the extrusion process itself, which squeezed through a cylinder by a specially designed volute [26] can increase the temperature above 100 °C. Indeed, many researchers have shown that extrusion is an efficient procedure for decreasing the trypsin inhibitory capacity of pulses like soybean without altering the amino acid composition, transforming soybean into a high-quality product [26].

For a better understanding of the potential mode of action of *Ulva* protease inhibitors, a kinetic study was performed in which commercial trypsin and chymotrypsin were exposed to *Ulva* extracts. Plant protease inhibitors are characterized by either reversible or irreversible mechanisms [28]. In the present study, kinetics studies revealed a potential reversible inhibition. According to the Lineweaver–Burk plots, both trypsin and chymotrypsin inactivation occurred by a mixed-type inhibition. This type of inhibition is characterized by the ability to bind not only to free enzymes but also to with the enzyme-substrate complexes [29]. In kinetics terms, mixed type inhibition causes changes that result in a progressive decrease in V_{max} when K_m increases occurs [30]. This type of inhibition cannot be reversed by increasing substrate concentration, given that the inhibitor cannot be displaced by the substrate. Therefore, the extent of the inhibition depends on the concentration of the inhibitor. In addition, differences in K_m indicated that *Ulva* protease inhibitors presented higher affinity for trypsin than for chymotrypsin.

Although plenty of literature regarding the mechanisms of plant protease inhibitors is available, there are no studies assessing the mode of action of seaweed protease inhibitors. Protease inhibitors in soybean and other seeds have been studied extensively. They are grouped into the Bowman–Birk and Kunitz families according to primary structure homology, the position of reactive sites, the number or location of disulfide bonds, and their ability to withstand thermal and acid processing [31]. According to the available literature, Kunitz proteinase inhibitors are usually 18–26 kDa proteins [32]. Overall, they are characterized by several Kunitz domains composed of approximately 60 amino acid residues, stabilized by three conserved disulfide bonds. This family mainly inhibits trypsin and weakly inhibits chymotrypsin [33] and is relatively heat- and acid-sensitive [15]. On the other hand, Bowman–Birk proteinase inhibitors (BBIs) are usually 6–9 kDa proteins with a polypeptide chain bridged by seven conserved disulfide bonds; they have independent sites for trypsin and for chymotrypsin, and they display similar inhibitory capacity for both proteases [31,34]. Disulfide bonds are fundamental for maintaining the structural stability of inhibitors [35]. Unfortunately, this work does not provide information on the structure of purified protease inhibitors of *Ulva*, their molecular weight, or their amino acid profile; however, the ability to inhibit mainly trypsin and also chymotrypsin, as well as the heat lability observed, suggest a certain similarity with the Kunitz type inhibitors family, although further studies are needed to ascertain this hypothesis.

4. Materials and Methods

4.1. *Ulva* Biomass

Ulva ohnoi biomass was ground, sieved (<100 μm), and kept at −20 °C until use. For inhibitory assays, *Ulva* biomass was separated in two different batches (100 g each), the first batch received no thermal treatment (CR-*Ulva*), and the second was heat-treated at 120 °C for 20 min (HT-*Ulva*). Aqueous extracts (0.1 g mL^{-1}) were prepared from CR-*Ulva* and HT-*Ulva*, homogenized in distilled water by shaking for 30 min at room temperature, and then for 24 h at 4 °C. The mixture was centrifuged for 20 min at 12,000 g and 4 °C. Supernatants were stored at 4 °C until used in posterior inhibitory assays.

4.2. Fish Enzyme Extracts

Juvenile specimens of gilthead seabream (*Sparus aurata*), seabass (*Dicentrarchus labrax*), and Senegalese sole (*Solea senegalensis*) were used as model aquaculture fish species. Nine fish of each species were anesthetized and sacrificed by severing their spine according to the requirements of the Council Directive 2010/63/UE. The abdomen was opened and the whole viscera were obtained. Intestines of each species were pooled (three pools including three intestines each, one per fish species), and manually homogenized in distilled water at 4 °C to a final concentration of 0.5 g mL^{-1}. Supernatants were obtained after centrifugation (12,000 rpm for 12 min at 4 °C) and stored at −20 °C until further use. The total soluble protein in the enzyme extracts was determined using bovine serum albumin as standard [36]. The total alkaline protease activity in the enzyme extracts was measured spectrophotometrically following the procedures described by Alarcón et al. [37], using 5 g L^{-1} casein in 50 mM Tris–HCl (pH 9.0) as substrate. One unit of total protease activity of activity (UA) was defined as the amount of enzyme that released 1 µg of tyrosine per min in the reaction mixture, considering an extinction coefficient for tyrosine of 0.008 μg^{-1} mL^{-1} cm^{-1}, measured spectrophotometrically at 280 nm.

4.3. Testing the Presence of Protease Inhibitors in *Ulva*

The inhibitory effects of CR-*Ulva* and HT-*Ulva* on the intestinal proteases of gilthead seabream, Senegalese sole and seabass were determined using a modification of the method described by Alarcón et al. [38]. This method is based on the measurement of the residual proteolytic activity after the preincubation of fish extracts with different volumes of CR-*Ulva* and HT-*Ulva* extracts providing a ratio mg *Ulva* per fish protease activity ranged from 0.0 mg *Ulva* UA^{-1} to 1.5 mg *Ulva* UA^{-1}. Enzyme inhibition was expressed as a percentage of protease inhibition after comparing with a control assay carried out without any *Ulva* extract. In addition, the amount of *Ulva* requested for 50% protease inhibition (IC_{50}) was estimated.

In order to visualize the effect of *Ulva* on active fish intestinal proteases, substrate-SDS-PAGE electrophoresis gels were performed. Intestinal extracts were preincubated for 60 min with different volumes of CR-*Ulva* or HT-*Ulva* extracts. Then, the samples were mixed (1:1) with SDS sample buffer (0.125 M Tris HCl, pH 6.8; 4% (*w*/*v*) SDS; 20% (*v*/*v*) glycerol; 0.04% (*w*/*v*) bromophenol blue) and SDS-PAGE was performed according to Laemmli, [39] using 11% polyacrylamide gels (100 V per gel, 45 min at 4 °C). Zymograms revealing protease active bands were made according to Alarcón et al. [37]. After electrophoresis, gels were washed with distilled water and incubated in 0.75% (*w*/*v*) casein solution prepared in 50 mM Tris–HCl buffer, pH 9.0, for 30 min at 4 °C. The gels were then incubated in the same solution for 90 min at 37 °C without agitation. Finally, the gels were washed and fixed in 12% TCA for 10 min to stop the reaction prior to staining with Coomassie Brilliant Blue R-250 in a solution of methanol–acetic acid–water for 12 h. Distaining was done using a methanol–acetic acid–water solution. Clear gel zones revealed the presence of active proteases with caseinolytic activity.

4.4. Effect of *Ulva* on Fish Digestive Proteases

The capacity of *Ulva* to inhibit the hydrolysis of casein by fish intestinal proteases was also assessed using an in vitro assay in the presence of different concentrations of crude (CR) and heat-treated (HT) algae biomass providing 0.5, 1.0 and 1.5 mg *Ulva* UA^{-1}.

The in vitro casein hydrolysis was simulated in 10 mL-jacketed reaction vessels connected to a circulating water bath at 37 °C, under continuous agitation by a magnetic stirrer. The temperature was selected in order to increase the activity of the enzymes for reducing the time requested for each analysis [40]. An amount of casein, providing 80 mg of crude protein per vessel, was suspended in 50 mM Tris HCl buffer pH 9.0. After 15 min stirring, the hydrolysis was started by the addition of the enzymatic extract providing 200 UA of total alkaline proteolytic activity [24]. The alkaline hydrolysis was maintained for 90 min, and samples of the reaction mixture (0, 15, 30, 60 and 90 min) were withdrawn. The products of the hydrolysis were separated by sodium dodecyl sulphate polyacrylamide gel electrophoresis (SDS-PAGE), and total amino acids released were also measured at each sampling time, in order to estimate the sequential degradation of casein [41]. All determinations were performed in triplicate. Blank assays with casein but without *Ulva* biomass were carried out for each fish species.

In order to visualize the casein hydrolysis of SDS-PAGE-separated casein fractions, electrophoresis gels were performed. The procedure was carried out as previously described. The rate of hydrolysis was expressed by a numerical value obtained considering both the percentage of reduction in optical density for each protein after the enzymatic hydrolysis and the relative proportion that such protein represented in the total proteins [38]. The value obtained was called the coefficient of protein degradation (CPD), and it was estimated using the following mathematical expression:

$$CDP = \sum_{i=1}^{n}\left[\frac{OD_i(t=0) - OD_i(t=90\text{min})}{OD_i(t=0)}\ x\ 100\right] x\ \frac{OD_i(t=0)}{\sum_{i=1}^{n} OD_i(t=0)}$$

where *i* are the proteins identified, OD_i is the optical density of the proteins, and *t* is the time of reaction.

In addition, the total released amino acids in each sampling time were also quantified at 340 nm in a spectrophotometer (Shimadzu UV-1800, Shimadzu, Kyoto, Japan), using L-leucine as standard [41]. The results were expressed as accumulated values of amino acid released during the enzymatic hydrolysis (g 100 g protein^{-1}).

4.5. Partial Characterization of *Ulva* Protease Inhibitors

4.5.1. Effect of Temperature on Protease Inhibitors

The effect of temperature on *Ulva* protease inhibitors was assessed by heating the aqueous *Ulva* extract (0.1 g mL^{-1}) at different temperatures (25, 40, 60, 80, 90, 100 °C) during 60 min and then immediately cooled in a water bath. Samples of *Ulva* from each temperature treatment were withdrawn at 5, 15, 30 and 60 min, and then preincubated with a solution of bovine trypsin (1 µg mL^{-1}. T8003 from Sigma Aldrich, SL. Saint Louis, MO, USA) during 60 min at room temperature at a ratio of 500 µg of *Ulva* per µg trypsin. After that, trypsin activity was assayed according to Erlanger et al. [42] using BAPNA (Nα-Benzoyl-DL-arginine 4-nitroanilide hydrochloride) as substrate. Enzyme inhibition was expressed as the percentage of trypsin inhibition after comparing with a control assay carried out without *Ulva*. SBTI was used as positive control of the inhibition assay.

In addition, the formation of proteinaceous enzyme-inhibitor complexes was determined by using substrate-SDS-PAGE electrophoresis gels. Samples were prepared by preincubating crude or heat-treated (100 °C, 5 min) *Ulva* extracts with a trypsin solution (1 µg mL^{-1}) at a ratio of 500 µg of *Ulva* per µg trypsin for 0, 30 and 60 min at room temperature. Samples were mixed (1:1) with SDS sample buffer (0.125 M Tris HCl, pH 6.8; 4% (*w*/*v*) SDS; 10% (*v*/*v*) β-mercaptoethanol; 20% (*v*/*v*) glycerol; 0.04% (*w*/*v*) bromophenol blue and SDS-PAGE was performed according to Laemmli [39] using 12% polyacrylamide gels (100 V per gel, 45 min, 4 °C). After electrophoresis, gels were washed with distilled water prior to staining with Coomassie Brilliant Blue R-250 in a methanol-acetic acid solution overnight. Finally, distaining was done with a methanol-acetic acid-water solution. In addition, 5 µL of a wide-range molecular weight marker (S-84445 SigmaMarker™, St. Louis, MO, USA) were included in each gel. The molecular marker consisted of 12 proteins ranging from 6.5 kDa (aprotinin, bovine lung) to 200 kDa (myosin, porcine heart).

4.5.2. Trypsin and Chymotrypsin Inhibition Kinetics

Inhibition kinetics were conducted according to Bijina et al. [30], with minor modifications, using trypsin and chymotrypsin from bovine pancreas (T8003 and C4129 from Sigma Aldrich, SL) and different concentrations of the synthetic substrate. An aliquot of 10 µL of each protease (1 mg mL^{-1}) was pre-incubated with different concentrations of *Ulva* (from 0 to 500 µg *Ulva* per µg trypsin) for 60 min. Later on, trypsin and chymotrypsin activities of the pre-incubated mixtures were assayed using various concentrations of BAPNA (Nα-Benzoyl-DL-arginine 4-nitroanilide hydrochloride) (from 0.10 to 0.75 mM) according to Erlanger et al. [42], or SAPNA (N-succinyl-(Ala)2-Pro-Phe-P-nitroanilide) (from 0.05 to 0.5 mM) according to DelMar et al. [43], respectively, in 50mM Tris-HCl, 10mM $CaCl_2$ buffer, pH 8.5.

The activity of the enzymatic reaction (v) based on the rate of change in absorbance (405 nm) of the reaction mixture was determined for each substrate concentration [S] assayed. Lineweaver–Burk curves, 1/v versus 1/[S], were plotted and the Michaelis constant (K*m*) and the maximum rate of reaction (V_{max}) were calculated for classifying the pattern of inhibition generated by the *Ulva* extract (competitive, uncompetitive or non-competitive).

4.6. Statistical Analysis

The results were expressed as mean ± standard deviation. In order to test data normality and variance homogeneity, the Kolmogorov–Smirnov test and Levene's F-test were used, respectively. Data with parametric distribution were analyzed using one-way analysis of variance (ANOVA), and the significant differences between treatments ($p < 0.05$) were determined using Tukey's multiple comparison test. Data with nonparametric distribution were analyzed by using Kruskal–Wallis test, and significant differences were determined using box-and-whisker plot graphs. All statistical analyses were performed using the Statgraphics Plus 4.0 (Rockville, MD, USA) software.

5. Conclusions

This work showed the presence of protease inhibitors in *Ulva* able to inhibit digestive proteases of commercial fish species. The inhibitory capacity was dose-dependent. From a physiological point of view, high dietary inclusion of crude *Ulva* would be requested to achieve higher inhibition values. The thermal treatment during feed processing is high enough to inactivate the inhibitors from *U. ohnoi*; hence, it can be used efficiently as potential sustainable ingredient for aquafeeds.

Author Contributions: A.J.V., M.I.S. and A.G. performed experiments and analyzed data. A.J.V. wrote the manuscript. F.J.A. and T.F.M. conceptualized, designed the research, revised and corrected the paper. All authors agreed with the final submitted version. All authors have read and agreed to the published version of the manuscript.

Funding: This research was funded by the projects SABANA (grant # 727874) from the European Union's Horizon 2020 Research and Innovation program, DORALGAE (RTI2018-096625-B-C31) from the Ministry of Sciences, Innovation and Universities (Spain) and the European Regional Development Fund (ERFD), and the knowledge transfer action grant GREEN4FEED # 5917 from the Ministry of Economy, Knowledge, Business and University of the Regional Government of Andalusia (Spain) and ERFD.

Acknowledgments: Authors thank IFAPA El Toruño center (Cádiz, Spain) for providing Senegalese sole specimens and *Ulva ohnoi* biomass. Gilthead seabream and seabass juveniles were kindly provided by Predomar (Almería, Spain).

Conflicts of Interest: The authors declare no conflict of interest.

References

1. Prabu, E.; Rajagopalsamy, C.B.T.; Ahilan, B.; Santhakumar, R.; Jeevagan, I.J.M.A.; Renuhadevi, M. An overview of anti-nutritional factors in fish feed ingredients and their effects in fish. *J. Aquac. Trop.* **2017**, *32*, 149.
2. Thakur, A.; Sharma, V.; Thakur, A. An overview of anti-nutritional factors in food. *Int. J. Chem. Stud.* **2019**, *7*, 2472–2479.
3. Glencross, B.D.; Baily, J.; Berntssen, M.H.; Hardy, R.; MacKenzie, S.; Tocher, D.R. Risk assessment of the use of alternative animal and plant raw material resources in aquaculture feeds. *Rev. Aquac.* **2019**, 1–56. [CrossRef]
4. Chakraborty, P.; Mallik, A.; Sarang, N.; Lingam, S.S. A review on alternative plant protein sources available for future sustainable aqua feed production. *Int. J. Chem. Stud.* **2019**, *7*, 1399–1404.
5. Hajra, A.; Mazumder, A.; Verma, A.; Ganguly, D.P.; Mohanty, B.P.; Sharma, A.P. Antinutritional factors in plant origin fish feed ingredients: The problems and probable remedies. *Adv. Fish Res.* **2013**, *5*, 193–202.
6. Mæhre, H.K. Seaweed Proteins—How to Get to Them? Effects of Processing on Nutritional Value, Bioaccessibility and Extractability. Ph.D. Thesis, The Arctic University of Norway, Tromsø, Norway, 2015.
7. Diken, G.; Demir, O.; Naz, M. The inhibitory effects of different diets on the protease activities of *Argyrosomus regius* (Pisces, Scianidae) larvae as a potential candidate species. *J. Appl. Anim. Res.* **2016**, *46*, 1–6. [CrossRef]
8. Vizcaíno, A.J.; Fumanal, M.; Sáez, M.I.; Martínez, T.F.; Moriñigo, M.A.; Fernández-Díaz, C.; Anguís, V.; Balebona, M.C.; Alarcón, F.J. Evaluation of *Ulva ohnoi* as functional dietary ingredient in juvenile Senegalese

sole (*Solea senegalensis*): Effects on the structure and functionality of the intestinal mucosa. *Algal Res.* **2019**, *42*, 101608. [CrossRef]
9. Guerreiro, I.; Magalhães, R.; Coutinho, F.; Couto, A.; Sousa, S.; Delerue-Matos, C.; Domingues, V.F.; Oliva-Teles, A.; Peres, H. Evaluation of the seaweeds *Chondrus crispus* and *Ulva lactuca* as functional ingredients in gilthead seabream (*Sparus aurata*). *J. Appl. Phycol.* **2019**, *31*, 2115–2124. [CrossRef]
10. Pereira, V.; Marques, A.; Gaivão, I.; Rego, A.; Abreu, H.; Pereira, R.; Santos, M.A.; Guilherme, S.; Pacheco, M. Marine macroalgae as a dietary source of genoprotection in gilthead seabream (*Sparus aurata*) against endogenous and exogenous challenges. *Comp. Biochem. Phys. C* **2019**, *219*, 12–24. [CrossRef]
11. Tapia-Paniagua, S.T.; Fumanal, M.; Anguis, V.; Fernandez-Diaz, C.; Alarcón, F.J.; Moriñigo, M.A.; Balebona, M.C. Modulation of intestinal microbiota in *Solea senegalensis* fed low dietary level of *Ulva ohnoi*. *Front. Microbiol.* **2019**, *10*, 171. [CrossRef]
12. Peixoto, M.J.; Magnoni, L.; Gonçalves, J.F.; Twijnstra, R.H.; Kijjoa, A.; Pereira, R.; Palstra, A.P.; Ozório, R.O. Effects of dietary supplementation of *Gracilaria* sp. extracts on fillet quality, oxidative stress, and immune responses in European seabass (*Dicentrarchus labrax*). *J. Appl. Phycol.* **2019**, *31*, 761–770. [CrossRef]
13. Silva, D.M.; Valente, L.M.P.; Sousa-Pinto, I.; Pereira, R.; Pires, M.A.; Seixas, F.; Rema, P. Evaluation of IMTA-produced seaweeds (*Gracilaria, Porphyra*, and *Ulva*) as dietary ingredients in Nile tilapia, *Oreochromis niloticus* L., juveniles. Effects on growth performance and gut histology. *J. Appl. Phycol.* **2015**, *27*, 1671–1680. [CrossRef]
14. Oliveira, M.N.; Ponte-Freitas, A.L.; Urano-Carvalho, A.F.; Taveres-Sampaio, T.M.; Farias, D.F.; Alves-Teixera, D.I.; Gouveia, S.T.; Gomes-Pereira, J.; Castro-Catanho de Sena, M.M. Nutritive and non-nutritive attributes of washed-up seaweeds from the coast of Ceará. *Food Chem.* **2009**, *11*, 254–259. [CrossRef]
15. Francis, G.; Makkar, H.P.; Becker, K. Antinutritional factors present in plant-derived alternate fish feed ingredients and their effects in fish. *Aquaculture* **2001**, *199*, 197–227. [CrossRef]
16. Le Bourvellec, C.; Renard, C.M. Interactions between polyphenols and macromolecules: Quantification methods and mechanisms. *Crit. Rev. Food Sci. Nutr.* **2012**, *52*, 213–248. [CrossRef] [PubMed]
17. Sáez, M.I.; Martinez, T.F.; Alarcón, F.J. Effect of the dietary of seaweeds on intestinal proteolytic activity of juvenile sea bream *Sparus Aurata*. *Int. Aquafeed* **2013**, *16*, 38–40.
18. Deivasigamani, B.; Subamanian, V. Applications of immunostimulants in aquaculture: A review. *Int. J. Curr. Microbiol. App. Sci.* **2016**, *5*, 447–453. [CrossRef]
19. Moutinho, S.; Linares, F.; Rodriguez, J.L.; Sousa, V.; Valente, L.M.P. Inclusion of 100% seaweed meal in diets for juvenile and on-growing life stages of Senegaleses sole (*Solea senegalensis*). *J. Appl. Phycol.* **2018**, *30*, 3589–3601. [CrossRef]
20. Haard, N.F.; Dimes, L.E.; Arndt, R.E.; Dong, F.M. Estimation of protein digestibility. IV. Digestive proteinases from the pyloric caeca of coho salmo (*Oncorhynchus kisutch*) fed diets containing soybean meal. *Comp. Biochem. Physiol.* **1996**, *115*, 533–540. [CrossRef]
21. Krogdahl, Å.; Penn, M.; Thorsen, J.; Refstie, S.; Bakke, A.M. Important antinutrients in plant feedstuffs for aquaculture: An update on recent findings regarding responses in salmonids. *Aquac. Res.* **2010**, *41*, 333–344. [CrossRef]
22. Magnoni, L.J.; Martos-Sitcha, J.A.; Queiroz, A.; Calduch-Giner, J.A.; Gonçalves, J.F.M.; Rocha, C.M.R.; Abreu, H.T.; Schrama, J.W.; Ozorio, R.O.A.; Pérez-Sánchez, J. Dietary supplementation of heat-treated *Gracilaria* and *Ulva* seaweeds enhanced acute hypoxia tolerance in gilthead sea bream (*Sparus aurata*). *Biol. Open* **2017**, *6*, 897–908. [CrossRef] [PubMed]
23. Bandara, T. Alternative feed ingredients in aquaculture: Opportunities and challenges. *J. Entomol. Zool. Stud.* **2018**, *6*, 3087–3094.
24. Vizcaíno, A.J.; Sáez, M.I.; Martínez, T.F.; Acién, F.G.; Alarcón, F.J. Differential hydrolysis of proteins of four microalgae by the digestive enzymes of gilthead sea bream and Senegalese sole. *Algal Res.* **2019**, *37*, 145–153. [CrossRef]
25. He, H.; Li, X.; Kong, X.; Hua, Y.; Chen, Y. Heat-induced inactivation mechanism of soybean Bowman-Birk inhibitors. *Food Chem* **2017**, *232*, 712–720. [CrossRef] [PubMed]
26. Vagadia, B.H.; Vanga, S.K.; Raghavan, V. Inactivation methods of soybean trypsin inhibitor—A review. *Trends Food Sci. Technol.* **2017**, *64*, 115–125. [CrossRef]

27. Maehre, H.K.; Edvinsen, G.K.; Eilertsen, K.E.; Elvevoll, E.O. Heat treatment increases the protein bioaccessibility in the red seaweed dulse (*Palmaria palmata*), but not in the brown seaweed winged kelp (*Alaria esculenta*). *J. Appl. Phycol.* **2016**, *28*, 581–590. [CrossRef]
28. Bijina, B.; Chellappan, S.; Krishna, J.G.; Basheer, S.M.; Elyas, K.K.; Bahkali, A.H.; Chandrasekaran, M. Protease inhibitor from *Moringa oleifera* with potential for use as therapeutic drug and as seafood preservative. *Saudi J. Biol. Sci.* **2011**, *18*, 273–281. [CrossRef]
29. Chang, T.S. An updated review of tyrosinase inhibitors. *Int. J. Mol. Sci.* **2009**, *10*, 2440–2475. [CrossRef]
30. Sharma, R. Enzyme inhibition: Mechanisms and scope. In *Enzyme inhibition and bioapplications*; InTech: Rijeka, Croatia, 2012; pp. 1–36.
31. Dantzger, M.; Vasconcelos, I.M.; Scorsato, V.; Aparicio, R.; Marangoni, S.; Macedo, M.L.R. Bowman—Birk proteinase inhibitor from *Clitoria fairchildiana* seeds: Isolation, biochemical properties and insecticidal potential. *Phytochemistry* **2015**, *118*, 224–235. [CrossRef]
32. Macedo, M.L.R.; Freire, M.G.M.; Cabrini, E.C.; Toyama, M.H.; Novello, J.C.; Marangoni, S. A trypsin inhibitor from *Peltophorum dubium* seeds active against pest proteases and its effect on the survival of *Anagasta kuehniella* (Lepidoptera:Pyralidae). *Biochim. Biophys. Gen. Subj.* **2003**, *1621*, 170–182. [CrossRef]
33. Xu, X.; Liu, J.; Wang, Y.; Si, Y.; Wang, X.; Wang, Z.; Zhang, Q.; Yu, H.; Wang, X. Kunitz-type serine protease inhibitor is a novel participator in anti-bacterial and anti-inflammatory responses in Japanese flounder (*Paralichthys olivaceus*). *Fish Shellfish Immunol.* **2018**, *80*, 22–30. [CrossRef]
34. Harry, J.B.; Steiner, R.F. A soybean proteinase inhibitor: Thermodynamic and kinetic parameters of association with enzymes. *Eur. J. Biochem.* **1970**, *16*, 174–179. [CrossRef] [PubMed]
35. Avilés-Gaxiola, S.; Chuck-Hernández, C.; del Refugio Rocha-Pizaña, M.; García-Lara, S.; López-Castillo, L.M.; Serna-Saldívar, S.O. Effect of thermal processing and reducing agents on trypsin inhibitor activity and functional properties of soybean and chickpea protein concentrates. *LWT Food Sci. Technol.* **2018**, *98*, 629–634. [CrossRef]
36. Bradford, M. A rapid and sensitive method for the quantitation of microgramquantities of protein utilizing the principle of protein-dye binding. *Anal. Biochem.* **1976**, *72*, 248–254. [CrossRef]
37. Alarcón, F.J.; Díaz, M.; Moyano, F.J.; Abellán, E. Characterization and functional properties of digestive proteases in two sparids; gilthead sea bream (*Sparus aurata*) and common dentex (*Dentex dentex*). *Fish Physiol. Biochem.* **1998**, *19*, 257–267. [CrossRef]
38. Alarcón, F.J.; García-Carreño, F.L.; Navarrete del Toro, M.A. Effect of plant protease inhibitors on digestive proteases in two fish species, *Lutjanus argentiventris* and *L. novemfasciatus*. *Fish Physiol. Biochem.* **2001**, *24*, 179–189. [CrossRef]
39. Laemmli, U.K. Cleavage of structural proteins during the assembly of the head of bacteriophage T4. *Nature* **1970**, *227*, 680–685. [CrossRef]
40. Hamdan, M.; Moyano, F.J.; Schuhardt, D. Optimization of a gastrointestinal model applicable to the evaluation of bioaccessibility in fish feeds. *J. Sci. Food Agric.* **2009**, *89*, 1195–1201. [CrossRef]
41. Church, F.C.; Swaisgood, H.E.; Porter, D.H.; Catignani, G. Spectrophotometric assay using o-phthaldehyde for determination of proteolysis in milk proteins. *J. Dairy Sci.* **1983**, *66*, 1219–1227. [CrossRef]
42. Erlanger, B.; Kokowsky, N.; Cohen, W. The preparation and properties of two new chromogenic substrates of trypsin. *Arch. Biochem. Biophys* **1961**, *95*, 271–278. [CrossRef]
43. DelMar, E.G.; Largman, C.; Brodrick, J.W.; Geokas, M.C. A sensitive new substrate for chymotrypsin. *Anal. Biochem.* **1979**, *99*, 316–320. [CrossRef]

Article

Secondary Metabolites with α-Glucosidase Inhibitory Activity from the Mangrove Fungus *Mycosphaerella* sp. SYSU-DZG01

Pei Qiu [1], Zhaoming Liu [1,2], Yan Chen [1], Runlin Cai [1], Guangying Chen [3,*] and Zhigang She [1,4,*]

1 School of Chemistry, Sun Yat-Sen University, Guangzhou 510275, China
2 State Key Laboratory of Applied Microbiology Southern China, Guangdong Institute of Microbiology, Guangdong Academy of Sciences, Guangzhou 510070, China
3 Key Laboratory of Tropical Medicinal Plant Chemistry of Ministry of Education, Hainan Normal University, Haikou 571158, China
4 South China Sea Bio-Resource Exploitation and Utilization Collaborative Innovation Center, Guangzhou 510006, China
* Correspondence: chgying123@163.com (G.C.); cesshzhg@mail.sysu.edu.cn (Z.S.); Tel.: +86-20-8411-3356 (Z.S.)

Received: 29 July 2019; Accepted: 18 August 2019; Published: 20 August 2019

Abstract: Four new metabolites, asperchalasine I (**1**), dibefurin B (**2**) and two epicoccine derivatives (**3** and **4**), together with seven known compounds (**5–11**) were isolated from a mangrove fungus *Mycosphaerella* sp. SYSU-DZG01. The structures of compounds **1–4** were established from extensive spectroscopic data and HRESIMS analysis. The absolute configuration of **1** was deduced by comparison of ECD data with that of a known structure. The stereostructures of **2–4** were further confirmed by single-crystal X-ray diffraction. Compounds **1**, **8** and **9** exhibited significant α-glucosidase inhibitory activity with IC_{50} values of 17.1, 26.7 and 15.7 μM, respectively. Compounds **1**, **4**, **6** and **8** showed antioxidant activity by scavenging DPPH· with EC_{50} values ranging from 16.3 to 85.8 μM.

Keywords: secondary metabolites; *Mycosphaerella* sp.; asperchalasine; α-glucosidase

---

1. Introduction

Diabetes mellitus (DM), a chronic metabolic disorder disease, is caused by the lack of insulin secretion (type Idiabetes mellitus) or insufficient insulin sensitivity (type IIdiabetes mellitus) [1,2], and the typical characteristic of the latter is post-prandial hyperglycemia. α-Glucosidase is a kind of membrane-bounded enzyme which is mainly found in intestinal epithelium cells and leads to the increase of blood glucose levels by hydrolyzing the glycosidic bonds of a polysaccharide [3–5]. As a result of that, α-Glucosidase inhibitors (AGIs), such as acarbose, miglitol and voglibose have become a widespread medical treatment in type IIdiabetes mellitus according to their glycemic control ability [6,7]. Nevertheless, existing AGIs often cause many side effects including abdominal pain, flatulence, diarrhea and other gastrointestinal disorders [8,9]. Hence, many natural medicine chemists were attracted to develop α-glucosidase inhibitors with lower toxicity and side effects for potential use. Some new α-glucosidase inhibitors have been researched like flavipesolides A–C [10], asperteretal E [11] and so on [12,13], and the discovery of better α-glucosidase inhibitors is still an urgent need.

Marine fungi are proved to be rich sources of structurally unique and bioactive secondary metabolites. *Mycosphaerella* sp., which contributes the largest genus of Ascomycota, is a common plant pathogen widely distributed in terrestrial plant and marine environment [14–16]. As part of our ongoing investigation on new secondary metabolites from marine fungi in the South China Sea [17–21], a mangrove fungus, *Mycosphaerella* sp. SYSU-DZG01, collected from the fruit of the mangrove plant *Bruguiera*, Hainan Dongzhai Harbor Mangrove Reserve attracted our attention because the EtOAc extract of the solid fermentation medium exhibited significant α-glucosidase inhibitory activity. Chemical investigation of the bioactive extract (Figure 1) lead to the discovery of four new metabolites, asperchalasine I (**1**), dibefurin B (**2**) and two epicoccine derivatives, (*R*)-9-((*R*)-10-hydroxyethyl)-7,9-dihydroisobenzofuran-1-ol (**3**), 2-methoxycarbonyl- 4,5,6-trihydroxy-3-methyl-benzaldehyde (**4**), together with seven known compounds, epicoccone B (**5**) [22], 1,3-dihydro-5-methoxy-7-methylisobenzofuran (**6**) [23], paeciloside A (**7**) [24], epicoccolide B (**8**) [25], asperchalasine A (**9**) [26], aspochalasin I (**10**) [27] and epicolactone (**11**) [28]. Their structures were established by extensive spectroscopic data and single-crystal X-ray diffraction analysis. Asperchalasine I possesses a distinct T-shaped skeleton containing one epicoccine moiety and one cytochalasan moiety. In bioactivity assays, compounds **1**, **8** and **9** exhibited α-glucosidase inhibitory activity and **1**, **4**, **6** and **8** showed antioxidant activity by scavenging DPPH·. Herein, the isolation, structure elucidation, α-glucosidase inhibitory and antioxidant activities of these compounds are reported.

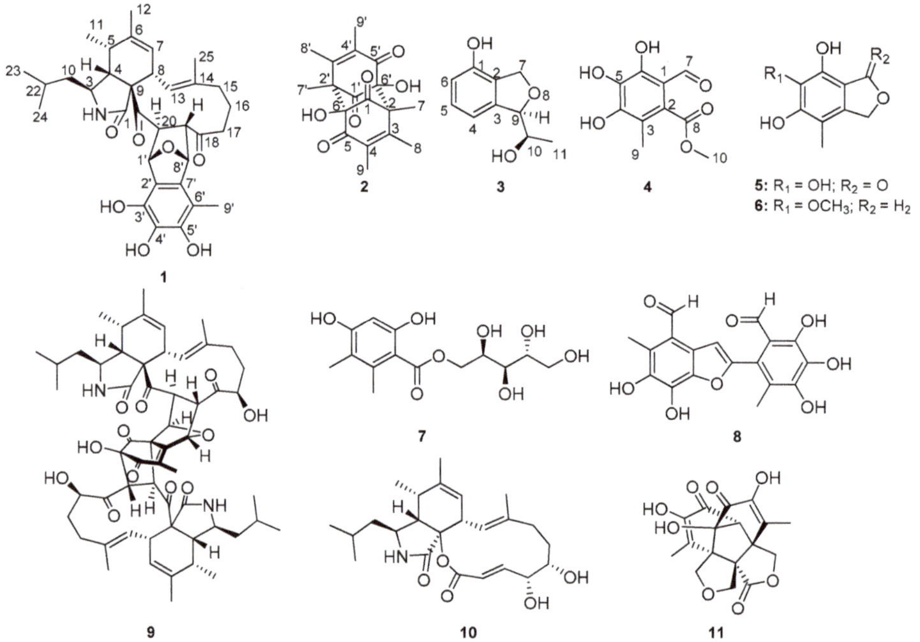

Figure 1. Chemical structures of **1**–**11**.

2. Results

2.1. Structure Elucidation

Asperchalasine I (**1**), the molecular formula, $C_{33}H_{41}O_7N$, was determined on the basis of the HRESIMS ion at *m*/*z* 562.2798 ($[M - H]^-$ calcd. for $C_{33}H_{40}O_7N$: 562.2799). As shown in Table 1, the ^1H NMR data indicated characteristic signals of two double-bond protons (δ_H 5.93 and 5.36), nine methine protons (δ_H 5.52, 5.06, 4.46, 3.76, 3.21, 3.05, 2.88, 2.54 and 1.62), four methylene protons (δ_H 2.09, 1.90, δ_H 2.01, 1.58, δ_H 1.35 and δ_H 1.21) and six methyl groups (δ_H 2.05, 1.78, 1.26, 1.17, 0.96 and 0.94). Subsequently, the ^{13}C NMR data showed the presence of 33 carbon signals, according to the DEPT and HSQC data, which were identified as three carbonyls (δ_C 211.6, 203.6 and 176.3), an aromatic ring (δ_C 141.3, 133.7, 132.6, 132.5, 123.7 and 111.9), two trisubstituted double bonds (δ_C 141.8, 137.4, 125.8 and 125.1), nine methines, four methylenes, six methyls and a quaternary carbon (δ_C 67.3). The ^1H-^1H COSY (Figure 2) correlations of H_3-23/H-22/H_2-10/H-3/H-4/H-5/H_3-11, H-7/H-8/H-13, H_2-15/H_2-16/H_2-17 and

H-19/H-20, together with the HMBC correlations from H-4 to C-1, C-9 and C-21, from H_3-12 to C-5, C-6 and C-7, from H-13 to C-25, from H_3-25 to C-15, from H_2-17 and H-19 to C-18, and from H-20 to C-21, suggested the presence of an cytochalasan moiety. Meanwhile, the epicoccine moiety was inferred by HMBC correlations from H-1′ to C-3′, from H-8′ to C-2′ and from H_3-9′ to C-5′, C-6′ and C-7′. The key linking relation of the cytochalasan moiety and epicoccine moiety through C-19/C-8′ and C-20/C-1′ C-C bonds was suggested from the ^{1}H-^{1}H COSY cross-peak of H-19/H-8′ and the HMBC correlation from H-1′ to C-20 and C-21. Moreover, the structure of **1** was further confirmed by a detailed comparison of the NMR data of **1** and asperchalasine B [26], which suggested that they shared the same skeleton. The upfield shifted of H_2-17 (δ_H 1.21) in **1** (δ_H 4.10 in asperchalasine B) and the disappearance of a methoxy group at C-4′ (δ_H 3.70, δ_C 61.0), suggesting the replacements of methine at C-17 and a methoxy group at C-4′ in asperchalasine B with a methylene and hydroxy group in **1**, respectively.

Table 1. ^{1}H (500 MHz) and ^{13}C (125 MHz) NMR data for **1** in $CDCl_3$.

Position	1	
	δ_H, mult (*J* in Hz)	δ_C, Type
1		176.3, C
2		
3	3.21, m	51.9, CH
4	3.05, dd (3.9, 5.0)	49.8, CH
5	2.54, m,	35.2, CH
6		141.8, C
7	5.36, s	125.1, CH
8	2.88, d (11.1)	43.3, CH
9		67.3, C
10	1.35, t	48.5, CH_2
11	1.26, d (7.3)	13.6, CH_3
12	1.78, s	20.1, CH_3
13	5.93, d (11.0)	125.8, CH
14		137.4, C
15	2.09, s	40.7, CH_2
	1.90, d (4.3)	
16	2.01, m	22.5, CH_2
	1.58, m	
17	1.21, m	35.1, CH_2
18		211.6, C
19	3.76, t (5.3)	56.9, CH
20	4.46, d (5.8)	57.3, CH
21		203.6, C
22	1.62, m	25.3, CH
23	0.94, dd (1.2, 6.5)	23.6, CH_3
24	0.96, dd (1.2, 6.5)	21.3, CH_3
25	1.17, s	14.2, CH_3
1′	5.06, s	80.7, CH
2′		123.7, C
3′		132.5, C
4′		133.7, C
5′		141.3, C
6′		111.9, C
7′		132.6, C
8′	5.52, d (4.9)	81.3, CH
9′	2.05, s	11.9, CH_3

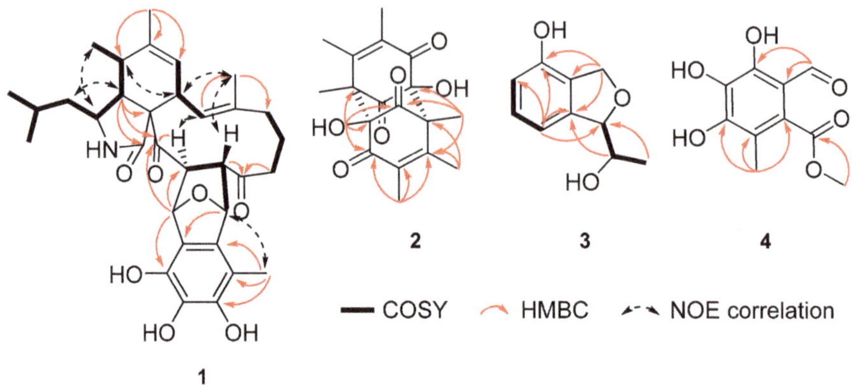

Figure 2. The key 2D NMR correlations of **1–4**.

Detailed analysis of NOESY data determined the relative configuration of chiral carbons in compound **1**. The NOESY correlations of H-4/H_2-10, H-3/H_3-11, H-5/H-8/H_3-25, H-13/H-20 and H-19/H_3-25 suggested that these protons were cofacial. Neither ^{1}H-^{1}H coupling nor a ^{1}H-^{1}H COSY correlation was observed between the protons of H-20 (δ_H 4.46, d, J = 5.8 Hz) and H-1′ (δ_H 5.06, s), which suggested that the dihedral angle of those protons was approximately 90° (Figure 3) [26]. The electron circular dichroism (ECD) spectrum (Figure S8) showed two positive Cotton effects (CE) at 228 nm and 306 nm, which were also consistent with those of asperchalasine B. Therefore, the absolute configuration of **1** was suggested to be 3*S*, 4*R*, 5*S*, 8*S*, 9*S*, 19*S*, 20*S*, 1′*S*, 8′*R*.

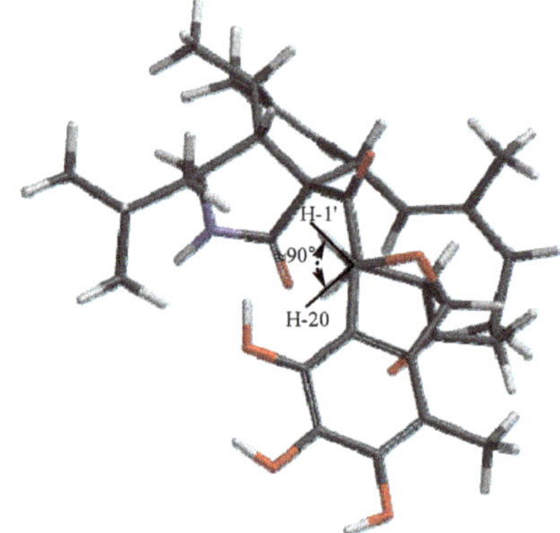

Figure 3. Molecular model of **1** (1′β, 8′ β-oxygen bridge).

Dibefurin B (**2**), a colorless monocrystal, was assigned a molecular formula of $C_{18}H_{20}O_6$ by the HRESIMS ion at *m/z* 331.1187 ($[M - H]^-$ calcd. for $C_{18}H_{19}O_6$: 331.1187). For the monomer, the ^{1}H, ^{13}C NMR (Table 2) and HSQC data of **2** revealed diagnostic signals for nine carbons, including two carbonyls (δ_C 200.6 and 192.6), two disubstituted olefin carbons (δ_C 157.6 and 131.4), an oxygen-bearing carbon [δ_H 6.67 (OH-6), δ_C 91.7 (C-6)], three methyls (δ_C 19.1, 12.2 and 12.1) and a quaternary carbon (δ_C 60.4). The HMBC correlations from H_3-7 to C-1, C-2, C-3 and C-6′, from H_3-8 to C-2, C-3 and C-4, from H_3-9 to C-3, C-4 and C-5, revealed that Me-7, Me-8 and Me-9 were connected at C-2, C-3 and C-4, respectively. Similarly, the hydroxyl was attached to C-6, as evidenced by the HMBC correlations from OH-6 to C-1, C-2′, C-5 and C-6. In view of the HRESIMS data and X-ray (Figure 4), it could confirm that compound **2** was a symmetric dimer and the absolute configuration of **2** was assigned as 2*S*, 6*R*, 2′*R*, 6′*S*.

Table 2. ^{1}H (400 MHz) and ^{13}C (100 MHz) NMR data for **2** in DMSO-d_6.

Position	**2**	
	δ_H, mult (*J* in Hz)	δ_C, Type
1 (1′)		200.6, C
2 (2′)		60.4, C
3 (3′)		157.6, C
4 (4′)		131.4, C
5 (5′)		192.6, C
6 (6′)		91.7, C
7 (7′)	1.21, s	12.1, CH_3
8 (8′)	2.02, s	19.1, CH_3
9 (9′)	1.73, s	12.2, CH_3

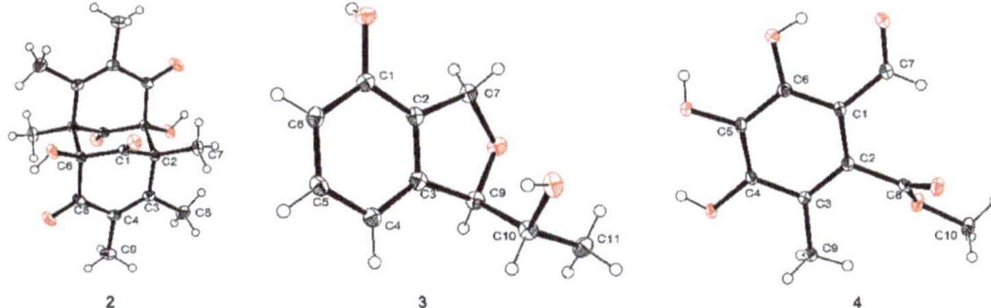

Figure 4. Single-crystal X-ray structures of **2**–**4**.

Compound **3** was purified as a colorless crystal whose molecular formula was deduced as $C_{10}H_{12}O_3$ based on HRESIMS data (179.0716 $[M - H]^-$, calcd. 179.0714). Analysis of the ^{1}H NMR spectrum of **3** (Table 3) displayed three aromatic proton resonances (δ_H 7.13, 6.82 and 6.69), an oxygenated methylene (δ_{Ha} 5.12, δ_{Hb} 5.02), two oxygenated methines (δ_H 5.06 and 3.97), and a methyl (δ_H 1.20). The ^{13}C spectrum revealed 10 signals, indicating an aromatic ring, a methylene, two methines and a methyl. In the ^{1}H-^{1}H COSY spectrum, the ortho-trisubstitution on the aromatic ring was confirmed by the cross-peaks of H-4/H-5/H-6. Moreover, the ^{1}H-^{1}H COSY spectrum showed correlations from H-10 to H-9 and H-11, and the chemical shift of C-10 (δ_C 70.6) showed the hydroxyl was attached to C-10. Subsequently, the HMBC correlations between H-6 and C-1/C-2 determined the linkage of 1-OH to C-1, and the correlations between H-7 and C-1/C-2/C-3, H-9 and C-2/C-3/C-10 established the presence of a phthalan ring. The same relative configuration of C-9 and C-10 was clearly deduced under the guidance of single-crystal X-ray (Figure 4). Hence, the absolute configuration of **3** was determined as 9*R*, 10*R*.

Compound **4** was deduced to have a molecular formula of $C_{10}H_{10}O_6$ from its HRESIMS spectrum with a deprotonated molecular ion at *m*/*z* 225.0407. The ^{1}H NMR (Table 3) in MeOH-d_4 showed three singlets at δ_H 9.71, 3.90 and 2.08, according to the ^{13}C NMR and HSQC data, which were attributed to an aldehyde group (δ_C 194.7), a methoxy group (δ_C 52.9) and a methyl (δ_C 12.4), respectively. In addition, resonances of a carbonyl and an aromatic ring were observed in the ^{13}C NMR data. In the HMBC spectrum, the correlations of H-9 to C-2, C-3 and C-4 supported the connection of Me-9 to C-3, the correlations of H-7 to C-1 and C-6 indicated the linkage of aldehyde group and C-1. Meanwhile, the carbonyl (δ_C 169.9, C-8) had the HMBC correlation from H-10 (δ_H 3.90), further indicated the presence of methyl ester. With the assistance of single-crystal X-ray (Figure 4), the structure of compound **4** was clearly confirmed.

Table 3. ^1H and ^{13}C NMR data for **3** and **4** in MeOH-d_4.

Position	3 [a]		4 [b]	
	δ_H, mult (J in Hz)	δ_C, Type	δ_H, mult (J in Hz)	δ_C, Type
1		152.8, C		112.2, C
2		127.3, C		117.4, C
3		142.1, C		129.9, C
4	6.82, d (7.4)	114.3, CH		152.5, C
5	7.13, t (7.7)	129.9, CH		134.1, C
6	6.69, d (7.9)	115.0, CH		151.2, C
7	5.12, dd (2.8, 11.8) 5.02, d (11.9)	72.4, CH_2	9.71, s	194.7, CH
8				169.9, C
9	5.06, d (3.3)	89.4, CH	2.08, s	12.4, CH_3
10	3.97, qd (3.9, 6.4)	70.6, CH	3.90, s	52.9, CH_3
11	1.20, d (6.4)	18.6, CH_3		

[a] ^1H and ^{13}C NMR recorded at 500 MHz and 125 MHz; [b] ^1H and ^{13}C NMR recorded at 400 MHz and 100 MHz.

2.2. Biological Evaluation

Compounds **1–11** were tested for their inhibitory effects against α-glucosidase, and antioxidant activity. As seen in Table 4, the results indicated that compounds **1**, **8** and **9** showed significant inhibitory effects against α-glucosidase with IC_{50} values of 17.1, 26.7 and 15.7 μM, respectively, which were better than the positive controls acarbose (610.2 μM) and 1-deoxynojirimycin (71.5 μM). Beyond that, all of the compounds were tested for their antioxidant activity based on DPPH· (2, 2-diphenyl-1-picrylhydrazyl radical) scavenging. The results showed the antioxidant activity of **8** was 89% at the concentration of 100 μM and compound **8** possessed more potent capacity than positive control ascorbic acid in scavenging DPPH· with an EC_{50} value of 16.3 μM. Compounds **1**, **4** and **6** also exhibited weak DPPH· scavenging activity with respective EC_{50} values of 77.8, 85.8 and 59.1 μM.

Table 4. The α-glucosidase inhibitory and antioxidant activities of compounds **1–11**.

Compounds	α-Glucosidase Inhibitory	Antioxidant	
	IC_{50} (μM)	% Inhibition (100 μM)	EC_{50} (μM)
1	17.1	56	77.8
2	>50	<50	-
3	>50	<50	-
4	>50	57	85.8
5	>50	<50	-
6	>50	65	59.1
7	>50	<50	-
8	26.7	89	16.3
9	15.7	<50	-
10	-	-	-
11	>50	<50	-
Acarbose [a]	610.2		
1-Deoxynojirimycin [a]	71.5		
Ascorbic acid [a]		92	22.4

- means no test; [a] positive control.

3. Experimental Section

3.1. General Experimental Procedures

UV data were measured on a UV-Vis-NIR spectrophotometer (Perkin Elmer, Waltham, UK). IR spectrum data were recorded using a Bruker Vector spectrophotometer 22. Melting points were tested on a Fisher-Johns hot-stage apparatus which were uncorrected. Optical rotations were recorded using an MCP300 (Anton Paar, Shanghai, China). HRESIMS data were conducted on an Ion Mobility-Q-TOF High-Resolution LC-MS (Synapt G2-Si, Waters). The ECD experiment data were measured with J-810 spectropolarimeter (JASCO, Tokyo, Japan). The NMR spectra were recorded on Bruker Avance spectrometer (Bruker, Beijing, China) (Compounds **1** and **3**: 500 MHz for ^{1}H and 125 MHz for ^{13}C, respectively; compounds **2** and **4**: 400 MHz for ^{1}H and 100 MHz for ^{13}C). Column chromatography (CC) was carried out on silica gel (200–300 mesh, Marine Chemical Factory, Qingdao, China) and sephadex LH-20 (Amersham Pharmacia, Piscataway, NJ, USA).

3.2. Fungal Materials

The fungus used in this research was isolated from the fruit of the marine mangrove plant *Bruguiera* collected in 2014 in Hainan Dongzhai Harbor Mangrove Reserve by using the standard protocol. The strain was identified as *Mycosphaerella* sp. (compared to no. KX067865.1) upon the analysis of ITS sequence data of the rDNA gene. The ITS sequence data obtained from the fungal strain has been submitted to GenBank with accession no. MN194208. A voucher strain was deposited in our laboratory.

3.3. Fermentation, Extraction and Isolation

The fungus *Mycosphaerella* sp. SYSU-DZG01 was grown on solid cultured medium in 100 × 1000 mL Erlenmeyer flasks at room temperature for 30 days under static conditions, each containing 80 g rice and 120 mL 0.3% saline. After incubation, the former was extracted with methanol twice and concentrated to yield 10.9 g of crude extract under reduced pressure. The crude extract was subjected to LC-HRESIMS analysis (Figure S29). Then, the residue was eluted by using gradient elution with petroleum ether/EtOAc from 9:1 to 0:10 (*v*/*v*) on silica gel CC to get ten fractions (Fr.1–Fr.10). Fr.2 (630 mg) was further eluted by silica gel CC using CH_2Cl_2/MeOH (40:1) to obtain Fr.2.1–Fr.2.3. Fr.2.3 (301 mg) was purified by Sephadex LH-20 CC and eluted with MeOH to obtain compound **2** (3.5 mg), **6** (11.1 mg) and **10** (3.6 mg). Fr.4 (217 mg) was applied to silica gel CC by CH_2Cl_2/MeOH (20:1) to obtain Fr.4.1–Fr.4.7. Fr.4.1 (8.1 mg) was further purified by Sephadex LH-20 CC using MeOH to obtain compound **1** (2.3 mg), **3** (2.2 mg), **8** (20.8 mg) and **9** (2.7 mg). Fr.5 (817 mg) was eluted (by CH_2Cl_2/MeOH, 25:3) to obtain Fr.5.1–Fr.5.5. Fr.5.1 (13.3 mg), Fr.5.2 (27.7 mg) and Fr.5.4 (10.9 mg) was purified by Sephadex LH-20 CC using CH_2Cl_2/MeOH (1:1) to yield compound **4** (4.3 mg), **5** (2.0 mg), **7** (3.7 mg) and **11** (2.7 mg).

Asperchalasine I (**1**): White powder; $[\alpha]_D^{25}$ = +61.4 (*c* 0.1, MeOH); UV (MeOH) λ_{max} (log ε): 206 (4.53) nm; IR (KBr) ν_{max} (cm^{-1}): 3369, 1691, 1440, 1384, 1201, 1120, 1053; HRESIMS *m*/*z* 562.2798 [M − H]$^-$ (calcd. for $C_{33}H_{40}O_7N$: 562.2799); ^{1}H and ^{13}C NMR data: see Table 1.

Dibefurin B(**2**): Colorless crystal; m.p. 67.8–69.2 °C; $[\alpha]_D^{25}$ = +0.3 (*c* 0.1, MeOH); UV (MeOH) λ_{max} (log ε): 237 (3.98) nm; IR (KBr) ν_{max} (cm^{-1}): 3448, 1747, 1664, 1645, 1238, 1037; HRESIMS *m*/*z* 331.1187 [M − H]$^-$ (calcd. for $C_{18}H_{19}O_6$, 331.1187); ^{1}H and ^{13}C NMR data: see Table 2.

Compound **3**: Colorless crystal; m.p. 89.8–91.9 °C; $[\alpha]_D^{25}$ = −37.1 (*c* 0.1, MeOH); UV (MeOH) λ_{max} (log ε) 204 (4.39), 269 (3.24) nm; IR (KBr) ν_{max} (cm^{-1}): 3261, 2887, 1601, 1471, 1297, 767, 706; HRESIMS *m*/*z* 179.0716 [M − H]$^-$ (calcd. for $C_{10}H_{12}O_3$, 179.0714; ^{1}H and ^{13}C NMR data: see Table 2.

Compound **4**: Colorless crystal; m.p. 95.9–97.8 °C; $[\alpha]_D^{25}$ = +3.4 (*c* 0.1, MeOH); UV (MeOH) λ_{max} (log ε) 241 (3.66), 305 (3.56) nm; IR (KBr) ν_{max} (cm^{-1}): 3375, 2962, 1711, 1641, 1261, 1150, 933; HRESIMS *m*/*z* 225.0407 [M − H]$^-$ (calcd. for $C_{10}H_{10}O_6$, 225.0407; ^{1}H and ^{13}C NMR data: see Table 2.

3.4. X-Ray Crystallographic Data

Colorless crystals of compounds **2–4** were obtained from MeOH-CH_2Cl_2 at room temperature by slow volatilization, and examined on an Agilent Xcalibur Nova single crystal diffractometer with Cu Kα radiation.

The crystallographic data for compound **2** has been deposited in the Cambridge Crystallographic Data Centre (CCDC number: 18022803)

Crystal data of **2**: $C_{18}H_{20}O_6$, Mr = 332.34, triclinic, a = 6.9740(4) Å, b = 7.9800(4) Å, c = 14.3659(6) Å, α = 101.148(4)°, β = 99.209(4)°, γ = 98.479(5)°, V = 761.05(7)Å^3; space group P-1, Z = 2, Dc = 1.450 g/cm^3, μ = 0.908 mm^{-1} and F(000) = 352.0; Crystal dimensions: 0.40 × 0.30 × 0.02 mm^3. Independent reflections: 4998 (R_{int} = 0.0269). The final R_1 was 0.0522, wR_2 = 0.1447 [$I > 2\sigma(I)$]. The goodness of fit on F^2 was 1.048.

The crystallographic data for compound **3** has been deposited in the Cambridge Crystallographic Data Centre (CCDC number: 18121203)

Crystal data of **3**: $C_{10}H_{12}O_3$, Mr = 180.07, monoclinic, a = 4.7697(1) Å, b = 11.1895(3) Å, c = 9.1541(3) Å, α = 90°, β = 93.829(3)°, γ = 90°, V = 487.47(2) Å^3; space group $P21$, flack 0.14(18), Z = 2, Dc = 1.350 g/cm^3, μ = 0.872 mm^{-1} and F(000) = 212.0; Crystal dimensions: 0.40 × 0.10 × 0.05 mm^3. Independent reflections: 7438 (R_{int} = 0.0620). The final R_1 was 0.0445, wR_2 = 0.1258 [$I > 2\sigma(I)$]. The goodness of fit on F^2 was 1.046.

The crystallographic data for compound **4** has been deposited in the Cambridge Crystallographic Data Centre (CCDC number: 18120705)

Crystal data of **4**: $C_{10}H_{10}O_6$, Mr = 226.04, orthorhombic, a = 15.9208(7) Å, b = 6.6849(3) Å, c = 18.5162(7) Å, α = 90°, β = 90°, γ = 90°, V = 1970.66(14) Å^3; space group $Pbca$, Z = 8, Dc = 1.525 g/cm^3, μ = 1.108 mm^{-1} and F(000) = 944.0; Crystal dimensions: 0.25 × 0.03 × 0.03 mm^3. Independent reflections: 3833 (R_{int} = 0.0496). The final R_1 was 0.0485, wR_2 = 0.1328 [$I > 2\sigma(I)$]. The goodness of fit on F^2 was 1.050.

3.5. Biological Assays

3.5.1. Inhibitory Activity of α-Glucosidase

The α-glucosidase inhibitory activity was assayed according to the reported method [29]. The inhibitory activity of α-glucosidase was tested in the 96-well plated with 100 mm PBS (KH_2PO_4-K_2HPO_4, pH 7.0) buffer solution each. Compounds **1–11**, acarbose and 1-deoxynojirimycin (positive control) were dissolved in DMSO, the substrate (p-nitrophenyl glycoside, 5 mM) were dissolved in PBS buffer solution and enzyme solutions (2.0 units/mL) were prepared. The assay was conducted in a 100 µL reaction system containing 20 µL enzyme stock solution, 69 µL PBS buffers and 1 µL of DMSO or testing materials. After 10 min incubation at 37 °C, 10 µL of the substrate was added and incubated for 20 min at 37 °C. The Absorbance which measured by a BIO-RAD (iMark) microplate reader at 405 nm was used to calculate the inhibitory activity according to the equation:

$$\eta\,(\%) = [(B - S)/B] \times 100\% \quad (1)$$

η (%) is the percentage of inhibition; B is the assay medium with DMSO; S is the assay medium with compound. The results of IC_{50} values were calculated by the nonlinear regression analysis. Acarbose and 1-deoxynojirimycin were used as positive controls.

3.5.2. Antioxidant Activity

The DPPH· scavenging was assayed according to the reported method [30]. The DPPH radical scavenging test was performed in 96-well microplates. Testing materials (compounds **1–11**) were added to 150 µL (0.16 mmol/L) DPPH solution in MeOH at a range of 50 µL solutions of different concentrations (2, 25, 50 and 100 μM). After 30 min, absorbance at 517 nm was measured and the percentage of activity was calculated. Ascorbic acid was used as a positive control.

4. Conclusions

In summary, four new metabolites, including one new asperchalasine I (**1**), dibefurin B (**2**), two epicoccine derivatives (**3**,**4**) and seven known compounds were isolated from the fungus *Mycosphaerella* sp. SYSU-DZG01. The structures of **1–11** were established by spectroscopic data and the absolute configuration of compounds **1–3** was determined in this research. Compound **1** possesses a unique T-shaped skeleton. All of the compounds were tested for their biological activities. Compounds **1**, **8** and **9** exhibited inhibitory effects against α-glucosidase with IC_{50} values of 17.1, 26.7 and 15.7 μM, respectively while compounds **1**, **4**, **6** and **8** showed antioxidant activity by scavenging DPPH· with EC_{50} values of 77.8, 85.8, 59.1 and 16.3 μM. These results suggested that the asperchalasine I may be a potential candidate for α-glucosidase inhibitor.

Supplementary Materials: The following are available online at http://www.mdpi.com/1660-3397/17/8/483/s1, Figure S1: ^{1}H NMR spectrum of compound **1** (500 MHz, $CDCl_3$), Figure S2: ^{13}C NMR spectrum of compound **1** (125 MHz, $CDCl_3$), Figure S3: DEPT 135, DEPT 90 and ^{13}C NMR spectrum of compound **1** (125 MHz, $CDCl_3$), Figure S4: ^{1}H-^{1}H COSY spectrum of compound **1** ($CDCl_3$), Figure S5: HSQC spectrum of compound **1** ($CDCl_3$), Figure S6: HMBC spectrum of compound **1** ($CDCl_3$), Figure S7: NOESY spectrum of compound **1** ($CDCl_3$), Figure S8: HRESIMS spectrum of compound **1**, Figure S9: Experiment ECD spectrum of **1**, Figure S10: ^{1}H NMR spectrum of compound **2** (400 MHz, DMSO-d_6), Figure S11: ^{13}C NMR spectrum of compound **2** (100 MHz, DMSO-d_6), Figure S12: ^{1}H-^{1}H COSY spectrum of compound **2** (DMSO-d_6), Figure S13: HSQC spectrum of compound **2** (DMSO-d_6), Figure S14: HMBC spectrum of compound **2** (DMSO-d_6), Figure S15: HRESIMS spectrum of compound **2**, Figure S16: ^{1}H NMR spectrum of compound **3** (500 MHz, MeOH-d_4), Figure S17: ^{13}C NMR spectrum of compound **3** (125 MHz, MeOH-d_4), Figure S18: ^{1}H-^{1}H COSY spectrum of compound **3** (MeOH-d_4), Figure S19: HSQC spectrum of compound **3** (MeOH-d_4), Figure S20: HMBC spectrum of compound **3** (MeOH-d_4), Figure S21: NOESY spectrum of compound **3** (MeOH-d_4), Figure S22: HRESIMS spectrum of compound **3**, Figure S23: ^{1}H NMR spectrum of compound **4** (400 MHz, MeOH-d_4), Figure S24: ^{13}C NMR spectrum of compound **4** (100 MHz, MeOH-d_4), Figure S25: ^{1}H-^{1}H COSY spectrum of compound **4** (MeOH-d_4), Figure S26: HSQC spectrum of compound **4** (MeOH-d_4), Figure S27: HMBC spectrum of compound **4** (MeOH-d_4), Figure S28: HRESIMS spectrum of compound **4**. Figure S29: The LC-HRESIMS analysis profile of crude extract.

Author Contributions: P.Q. contributed to isolation, structure elucidation and wrote the paper; Z.L. contributed to the analysis of the NMR data and structure elucidation. Y.C. contributed to the analysis of the NMR data and biological tests. R.C. contributed to the spectral analysis. Z.S. and G.C. guided the whole experiment and revised the manuscript.

Funding: This research was funded by the Guangdong Special Fund for Marine Economic Development (GDME-2018C004), Guangdong MEPP Fund (GDOE-2019A21), the National Natural Science Foundation of China (2187713, 21472251), the Key Project of Natural Science Foundation of Guangdong Province (2016A040403091), the Special Promotion Program for Guangdong Provincial Ocean and Fishery Technology (A201701C06) and the open foundation of Key Laboratory of Tropical Medicinal Resource Chemistry of Ministry of Education (rdyw2018001).

Acknowledgments: We thank all the funding assistance mentioned above for their generous support and the Sun Yat-Sen University Instrumental Analysis and Research Center.

Conflicts of Interest: The authors declare no conflict of interest.

References

1. Wresdiyati, T.; Sa'Diah, S.; Winarto, A.; Febriyani, V. Alpha-Glucosidase Inhibition and Hypoglycemic Activities of Sweitenia mahagoni Seed Extract. *HAYATI J. Biosci.* **2015**, *22*, 73–78. [CrossRef]
2. Sekar, V.; Chakraborty, S.; Mani, S.; Sali, V.K.; Vasanthi, H.R. Mangiferin from Mangifera indica fruits reduces post-prandial glucose level by inhibiting α-glucosidase and α-amylase activity. *S. Afr. J. Bot.* **2019**, *120*, 129–134. [CrossRef]
3. Wang, H.; Du, Y.J.; Song, H.C. α-Glucosidase and α-amylase inhibitory activities of guava leaves. *Food Chem.* **2010**, *123*, 6–13. [CrossRef]

4. Tang, H.; Ma, F.; Zhao, D.; Xue, Z. Exploring the effect of salvianolic acid C on α-glucosidase: Inhibition kinetics, interaction mechanism and molecular modelling methods. *Process. Biochem.* **2019**, *78*, 178–188. [CrossRef]
5. Indrianingsih, A.W.; Tachibana, S. α-Glucosidase inhibitor produced by an endophytic fungus, *Xylariaceae* sp. QGS 01 from *Quercus gilva* Blume. *Food Sci. Hum. Wellness* **2017**, *6*, 88–95. [CrossRef]
6. Yin, Z.; Zhang, W.; Feng, F.; Zhang, Y.; Kang, W. α-Glucosidase inhibitors isolated from medicinal plants. *Food Sci. Hum. Wellness* **2014**, *3*, 136–174. [CrossRef]
7. Chen, J.; Zhang, X.; Huo, D.; Cao, C.; Li, Y.; Liang, Y.; Li, B.; Li, L. Preliminary characterization, antioxidant and α-glucosidase inhibitory activities of polysaccharides from *Mallotus furetianus*. *Carbohydr. Polym.* **2019**, *215*, 307–315. [CrossRef] [PubMed]
8. Lin, M.Z.; Chai, W.M.; Zheng, Y.L.; Huang, Q.; Ou-Yang, C. Inhibitory kinetics and mechanism of rifampicin on α-glucosidase: Insights from spectroscopic and molecular docking analyses. *Int. J. Biol. Macromol.* **2019**, *122*, 1244–1252. [CrossRef] [PubMed]
9. Wei, M.; Chai, W.M.; Yang, Q.; Wang, R.; Peng, Y. Novel Insights into the Inhibitory Effect and Mechanism of Proanthocyanidins from *Pyracantha fortuneana* Fruit on α-Glucosidase. *J. Food Sci.* **2017**, *82*, 2260–2268. [CrossRef]
10. Sun, Y.; Liu, J.; Li, L.; Gong, C.; Wang, S.; Yang, F.; Hua, H.; Lin, H. New butenolide derivatives from the marine sponge-derived fungus *Aspergillus terreus*. *Bioorgan. Med. Chem. Lett.* **2018**, *28*, 315–318. [CrossRef]
11. Liu, H.; Chen, Z.; Zhu, G.; Wang, L.; Du, Y.; Wang, Y.; Zhu, W. Phenolic polyketides from the marine alga-derived *Streptomyces* sp. OUCMDZ-3434. *Tetrahedron* **2017**, *73*, 5451–5455. [CrossRef]
12. Rizvi, T.S.; Hussain, I.; Ali, L.; Mabood, F.; Khan, A.L.; Shujah, S.; Rehman, N.U.; Al-Harrasi, A.; Hussain, J.; Khan, A.; et al. New gorgonane sesquiterpenoid from *Teucrium mascatense* Boiss, as α-glucosidase inhibitor. *S. Afr. J. Bot.* **2019**, *124*, 218–222. [CrossRef]
13. Wang, C.; Guo, L.; Hao, J.; Wang, L.; Zhu, W. α-Glucosidase Inhibitors from the Marine-Derived Fungus *Aspergillus flavipes* HN4-13. *J. Nat. Prod.* **2016**, *79*, 2977–2981. [CrossRef] [PubMed]
14. Lee, J.; Lee, J.; Kim, G.J.; Yang, I.; Wang, W.; Nam, J.W.; Choi, H.; Nam, S.J.; Kang, H. Mycousfurans A and B, Antibacterial Usnic Acid Congeners from the Fungus *Mycosphaerella* sp., Isolated from a Marine Sediment. *Mar. Drugs* **2019**, *17*, 422. [CrossRef] [PubMed]
15. Otálvaro, F.; Nanclares, J.; Vásquez, L.E.; Quiñones, W.; Echeverri, F.; Arango, R.; Schneider, B. Phenalenone-Type compounds from *Musa acuminata* var. "Yangambi km 5" (AAA) and Their Activity against *Mycosphaerella fijiensis*. *J. Nat. Prod.* **2007**, *70*, 887–890. [CrossRef] [PubMed]
16. Assante, A.; Camarda, L.; Merlini, L.; Nasini, G. Secondary metabolites from *Mycosphaerella ligulicola*. *Phytochemistry* **1981**, *20*, 1955–1957. [CrossRef]
17. Huang, H.; Feng, X.; Xiao, Z.; Liu, L.; Li, H.; Ma, L.; Lu, Y.; Ju, J.; She, Z.; Lin, Y. Azaphilones and *p*-Terphenyls from the Mangrove Endophytic Fungus *Penicillium chermesinum* (ZH4-E2) Isolated from the South China Sea. *J. Nat. Prod.* **2011**, *74*, 997–1002. [CrossRef]
18. Liu, Y.; Yang, Q.; Xia, G.; Huang, H.; Li, H.; Ma, L.; Lu, Y.; He, L.; Xia, X.; She, Z. Polyketides with α-Glucosidase Inhibitory Activity from a Mangrove Endophytic Fungus, *Penicillium* sp. HN29-3B1. *J. Nat. Prod.* **2015**, *78*, 1816–1822. [CrossRef]
19. Liu, Z.; Chen, S.; Qiu, P.; Tan, C.; Long, Y.; Lu, Y.; She, Z. (+)- and (−)-Ascomlactone A: A pair of novel dimeric polyketides from a mangrove endophytic fungus *Ascomycota* sp. SK2YWS-L. *Org. Biomol. Chem.* **2017**, *15*, 10276–10280. [CrossRef] [PubMed]
20. Cui, H.; Liu, Y.; Nie, Y.; Liu, Z.; Chen, S.; Zhang, Z.; Lu, Y.; He, L.; Huang, X.; She, Z. Polyketides from the Mangrove-Derived Endophytic Fungus *Nectria* sp. HN001 and Their α–Glucosidase Inhibitory Activity. *Mar. Drugs* **2016**, *14*, 86. [CrossRef] [PubMed]
21. Liu, Y.; Xia, G.; Li, H.; Ma, L.; Ding, B.; Lu, Y.; He, L.; Xia, X.; She, Z. Vermistatin Derivatives with α-Glucosidase Inhibitory Activity from the Mangrove Endophytic Fungus *Penicillium* sp. HN29-3B1. *Planta Med.* **2014**, *80*, 912–917. [CrossRef] [PubMed]
22. Kemami Wangun, H.V.; Ishida, K.; Hertweck, C. Epicoccalone, a Coumarin-Type Chymotrypsin Inhibitor, and Isobenzofuran Congeners from an *Epicoccum* sp. Associated with a Tree Fungus. *Eur. J. Org. Chem.* **2008**, *22*, 3781–3784. [CrossRef]
23. Nam, H.L.; James, B.G.; Donald, T.W. Isolation of Chromanone and Isobenzofuran Derivatives from a Fungicolous Isolate of *Epicoccum purpurascens*. *Bull. Korean Chem. Soc.* **2007**, *28*, 877–879.

24. Ferdinand, M.T.; Timothee, J.N.K.; Birger, D.; Clovis, D.M.; Hartmut, L. Paeciloside A, a new antimicrobial and cytotoxic polyketide from *Paecilomyces* sp. Strain CAFT156. *Planta Med.* **2012**, *78*, 1020–1023.
25. Talontsi, F.M.; Dittrich, B.; Schüffler, A.; Sun, H.; Laatsch, H. Epicoccolides: Antimicrobial and Antifungal Polyketides from an Endophytic Fungus *Epicoccum* sp. Associated with *Theobroma cacao*. *Eur. J. Org. Chem.* **2013**, *15*, 3174–3180. [CrossRef]
26. Zhu, H.; Chen, C.; Xue, Y.; Tong, Q.; Li, X.; Chen, X.; Wang, J.; Yao, G.; Luo, Z.; Zhang, Y. Asperchalasine A, a Cytochalasan Dimer with an Unprecedented Decacyclic Ring System, from *Aspergillus flavipes*. *Angew. Chem. Int. Ed.* **2015**, *54*, 13374–13378. [CrossRef] [PubMed]
27. Zhou, G.X.; Wijeratne, E.M.K.; Bigelow, D.; Pierson, L.S.; Vanetten, H.D.; Gunatilaka, A.A.L. Aspochalasins I, J, and K: Three New Cytotoxic Cytochalasans of *Aspergillus flavipes* from the Rhizosphere of *Ericameria laricifolia* of the Sonoran Desert. *J. Nat. Prod.* **2004**, *67*, 328–332. [CrossRef]
28. Da Silva Araújo, F.D.; de Lima Fávaro, L.C.; Araújo, W.L.; de Oliveira, F.L.; Aparicio, R.; Marsaioli, A.J. Epicolactone—Natural Product Isolated from the Sugarcane Endophytic Fungus *Epicoccum nigrum*. *Eur. J. Org. Chem.* **2012**, *27*, 5225–5230. [CrossRef]
29. Cai, R.; Wu, Y.; Chen, S.; Cui, H.; Liu, Z.; Li, C.; She, Z. Peniisocoumarins A–J: Isocoumarins from *Penicillium commune* QQF-3, an Endophytic Fungus of the Mangrove Plant *Kandelia candel*. *J. Nat. Prod.* **2018**, *81*, 1376–1383. [CrossRef]
30. Tan, C.; Liu, Z.; Chen, S.; Huang, X.; Cui, H.; Long, Y.; Lu, Y.; She, Z. Antioxidative Polyketones from the Mangrove-Derived Fungus *Ascomycota* sp. SK2YWS-L. *Sci. Rep.* **2016**, *6*, 36609. [CrossRef] [PubMed]

Article

Dual BACE1 and Cholinesterase Inhibitory Effects of Phlorotannins from *Ecklonia cava*—An In Vitro and in Silico Study

Jinhyuk Lee [1,2] and Mira Jun [3,4,5,*]

1 Korean Bioinformation Center, Korea Research Institute of Bioscience and Biotechnology (KRIBB), 125, Gwahak-ro, Yuseong-gu, Daejeon 34141, Korea; jinhyuk@kribb.re.kr
2 Department of Bioinformatics, KIRBB School of Bioscience, Korea University of Sciences and Technology, 217 Gajung-ro, Yuseong-gu, Daejeon 34113, Korea
3 Department of Food Science and Nutrition, College of Health Sciences, Dong-A University, 37, Nakdong-daero 550 beon-gil, Saha-gu, Busan 49315, Korea
4 Center for Silver-Targeted Biomaterials, Brain Busan 21 Plus Program, Graduate School, Dong-A University, Nakdong-daero 550 beon-gil, Saha-gu, Busan 49315, Korea
5 Institute of Convergence Bio-Health, Dong-A University, Busan 49315, Korea
* Correspondence: mjun@dau.ac.kr; Tel.: +82-51-200-7323

Received: 8 January 2019; Accepted: 28 January 2019; Published: 1 February 2019

Abstract: Alzheimer's disease (AD) is one of the most common neurodegenerative diseases with a multifactorial nature. β-Secretase (BACE1) and acetylcholinesterase (AChE), which are required for the production of neurotoxic β-amyloid (Aβ) and the promotion of Aβ fibril formation, respectively, are considered as prime therapeutic targets for AD. In our efforts towards the development of potent multi-target, directed agents for AD treatment, major phlorotannins such as eckol, dieckol, and 8,8′-bieckol from *Ecklonia cava* (*E. cava*) were evaluated. Based on the in vitro study, all tested compounds showed potent inhibitory effects on BACE1 and AChE. In particular, 8,8′-bieckol demonstrated the best inhibitory effect against BACE1 and AChE, with IC_{50} values of 1.62 ± 0.14 and 4.59 ± 0.32 μM, respectively. Overall, kinetic studies demonstrated that all the tested compounds acted as dual BACE1 and AChE inhibitors in a non-competitive or competitive fashion, respectively. In silico docking analysis exhibited that the lowest binding energies of all compounds were negative, and specifically different residues of each target enzyme interacted with hydroxyl groups of phlorotannins. The present study suggested that major phlorotannins derived from *E. cava* possess significant potential as drug candidates for therapeutic agents against AD.

Keywords: Alzheimer's disease; BACE1; acetylcholinesterase; in silico docking; phlorotannins

1. Introduction

Alzheimer's disease (AD) is a progressive and irreversible neurodegenerative disorder with characteristic features of cognitive dysfunction, memory impairment, and behavior disturbances. The neuropathological hallmarks of AD patients are the presence of extracellular deposits of amyloid plaques and intracellular filamentous neurofibrillary tangles in the brain [1]. Amyloid plaques and neurofibrillary tangles are aggregates of amyloid-β peptide (Aβ) and hyperphosphorylated tau protein, respectively. In recent years, the "amyloid hypothesis" has arisen as the major pathological mechanism in AD, and the evidence from transgenic mice models revealed that Aβ triggered tau phosphorylation and neurofibrillary tangles formation [2].

Aβ is generated by the sequential proteolytic cleavage of two aspartic proteases, β- and γ–secretase, in the amyloidogenic pathway. β-Secretase (BACE1) initially cleaves amyloid precursor protein (APP) at the N-terminus of the Aβ peptide domain, which is followed by the cleavage of γ–secretase in the transmembrane region of APP, leading to the production of Aβ peptide [3]. Therefore, these secretases have been suggested as potential targets to hinder Aβ formation and thereby delay or stop the progression of AD. γ-Secretase inhibitors have shown severe toxicity problems because of on-target interference with Notch signaling, and genetic deletion of catalytically active subunit presinilin-1 (PS-1) was found to be lethal in embryonic mice [4–6]. In contrast, BACE1 inhibition powerfully lowers the level of Aβ in the central nervous system (CNS) of both transgenic mice models and AD patients [7–9]. Recent studies have exhibited few mechanism-based side effects of BACE1 inhibition with chronic administration in animal models, but these are relatively weak and mild compared to γ-secretase-inhibitor-induced deficits [10–12]. Unlike BACE1, which leads to the formation of Aβ, α-secretase acts within the Aβ domain to preclude the Aβ generation in the non-amyloidogenic pathway. The α-secretase and BACE1 compete for the same APP substrate, with an increase in one cleavage event leading to a decrease in the other. As BACE1 initiates the amyloidogenic pathway and is putatively rate-limiting, it is a critical target for lowering cerebral Aβ levels in the treatment and/or prevention of AD.

In the past, it has been found that acetylcholinesterase (AChE) is involved in the degradation of the neurotransmitter. The observation of a significant loss of cholinergic neurons in AD patients is the major correlate of cognitive impairment. Cholinesterase inhibitors can increase acetylcholine (ACh) levels in the synaptic cleft and partially ameliorate cognitive symptoms for patients with mild to severe AD [13]. Recent findings supported that AChE is associated predominantly with pre-amyloid diffuse deposits, amyloid cores of mature amyloid plaques, and cerebral blood vessels in an AD patient brain. In addition, it triggers the Aβ fibrillogenesis via the formation of stable Aβ– AChE complexes [14,15]. Neurons treated with these complexes exhibited a disrupted neurite network compared to neurons treated with Aβ alone [16]. Based on these findings, the suppression of both enzymes is a very desirable feature of AD therapy.

Current AD therapies are mainly palliative and temporarily slow cognitive decline, and treatments based on the underlying pathologic mechanisms of AD are totally limited [17]. Several therapeutic approaches have recently revealed promising results in clinical trials, such as BACE1 and γ-secretase inhibitors, inhibition of Aβ plaque formation, passive Aβ immunotherapy, etc. However, the clinical use of these agents needs further careful assessment of their effectiveness on cognitive decline and their adverse effects [18]. Another strategy for AD therapy is the use of natural products, which are more effective, safer and have fewer adverse effects than synthesized drugs [19]. Neuroprotective natural compounds such as (-)-epigallocatechin-3-gallate (EGCG) from green tea, resveratrol from grape, curcumin from tumeric, and quercetin from apples revealed significant therapeutic potential toward the amelioration and prevention of AD [20].

Marine organisms are a rich source of several natural molecules, including polyphenol, polysaccharide, sterol, and peptide, which have many biological properties such as antioxidant, anti-inflammatory, anti-hypertensive, anti-obesity, anti-diabetes, and anti-cancer effects [21–25]. *Ecklonia cava* (*E. cava*) is an edible brown seaweed which is distributed in Japan and the southern coast of Korea, and it is recognized as a rich source of bioactive derivatives, containing 3.1% crude phlorotannins [26,27]. Phlorotannins are unique polyphenolic compounds containing a dibenzo[1,4]dioxin element as the core structure not found in terrestrial plants. The compounds consist of phloroglucinol units linked to each other in several ways. Based on the type of linkage, phlorotannins are classified into four subgroups: eckols (phlorotannins with a dibenzodioxin linkage), fuhalols and phlorethols (with an ether linkage), fucols (with a phenyl linkage), and fucophloroethols (with an ether and phenyl linkage) [27].

Recently, it has been reported that phlorotannins possess various bioactivities such as antioxidant, antidiabetic, anti-hypertensive, anti-human-immunodeficiency-virus type-1 (HIV-1), and radioprotective activities [28–32]. Regarding the study of neuroprotective effects, phlorotannins-rich *E. cava* extract ameliorated the Aβ formation by modulating α- and γ-secretase expression and inhibiting Aβ-induced neurotoxicity [19,33]. In our previous study, three major phlorotannins of E. cava—eckol, dieckol, and 8,8′-bieckol—exhibited anti-apoptotic and anti-neuroinflammatory properties against Aβ-induced cellular damage, which led to our interest in the study of phlorotannins-mediated suppression of related enzymes in Aβ production and aggregation [34]. Therefore, the purpose of the present study is to evaluate the inhibitory effects of these compounds against both BACE1 and AChE through in vitro and in silico approaches.

2. Results

2.1. In Vitro Inhibitory Study of Phlorotannins on BACE1 and AChE

The chemical structures of eckol [4-(3,5-dihydroxyphenoxy)dibenzo-p-dioxin-1,3,6,8-tetrol], dieckol [4-[4-[6-(3,5-dihydroxyphenoxy)-4,7,9-trihydroxydibenzo-p-dioxin-2-yl]oxy-3,5-dihydroxyp henoxy]dibenzo-p-dioxin-1,3,6,8-tetrol], and 8,8′-bieckol [9-(3,5-dihydroxyphenoxy)-2-[9-(3,5-dihy droxyphenoxy)- 1,3,6,8-tetrahydroxydibenzo-p-dioxin-2-yl]dibenzo-p-dioxin-1,3,6,8-tetrol] were shown in Figure 1. Eckol is a trimer of phloroglucinol (1,3,5-trihydroxybenzene) units, while dieckol and 8,8′-bieckol are hexamer. As presented in Table 1 and Figure 2, 8,8′-bieckol exhibited the strongest BACE1 inhibition (IC_{50}, 1.62 ± 0.14 μM), followed by dieckol (IC_{50}, 2.34 ± 0.10 μM) and eckol (IC_{50}, 7.67 ± 0.71 μM). Interestingly, the IC_{50} value of all of the tested compounds were lower than that of resveratrol (IC_{50}, 14.89 ± 0.54 μM), which is a well-known BACE1 inhibitor that was used as a positive control.

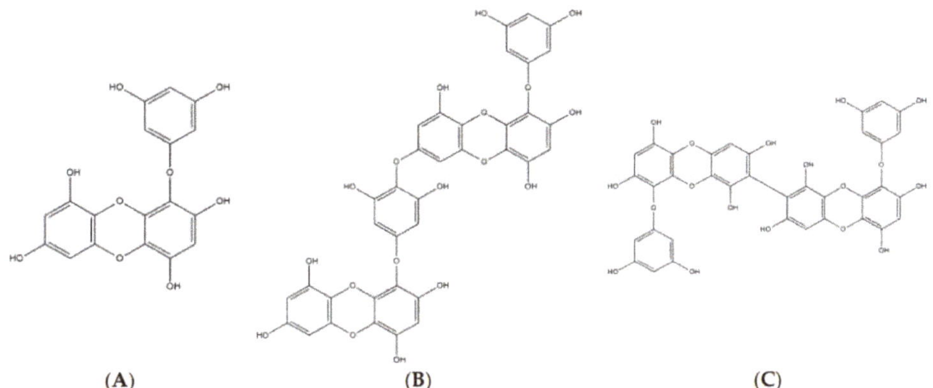

Figure 1. The chemical structures of (**A**) eckol, (**B**) dieckol, and (**C**) 8,8′-bieckol.

Table 1. Inhibitory activities of phlorotannins on BACE1 and AChE.

Compounds	IC_{50} (mean ± SD, μM) [a]		K_i value (μM) [d]		Inhibition Mode [e]	
	BACE1	AChE	BACE1	AChE	BACE1	AChE
Eckol	7.67 ± 0.71	10.03 ± 0.94	31.2	37.3	Non-competitive	Competitive
Dieckol	2.34 ± 0.10	5.69 ± 0.42	20.1	12.3	Non-competitive	Competitive
8,8′-Bieckol	1.62 ± 0.14	4.59 ± 0.32	13.9	11.4	Non-competitive	Competitive
Resveratrol [b]	14.89 ± 0.54	-	-	-	-	-
Galantamine [c]	-	0.99 ± 0.07	-	-	-	-

[a] The IC_{50} values (μM) were calculated from a log dose inhibition curve and expressed as the mean ± standard deviation (SD). All assays were performed in three independent experiments. DMSO was used as a negative control in the BACE1 and AChE assays. [b] Resveratrol and [c] galantamine were used as positive controls in the BACE1 and AChE assays, respectively. [d] Inhibition constant (K_i) and [e] inhibition mode were determined using Dixon plot and Lineweaver–Burk plot, respectively.

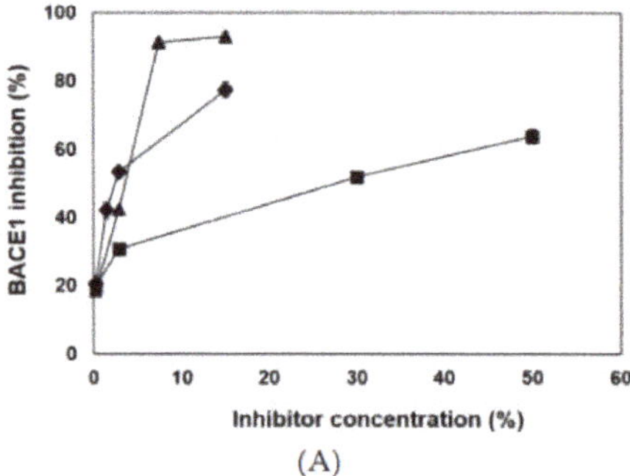

(A)

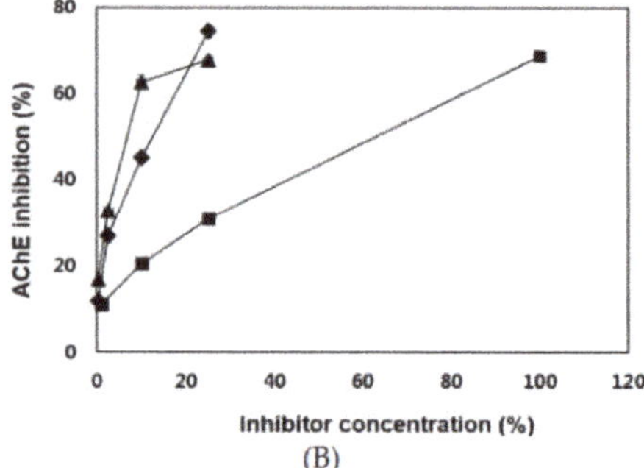

(B)

Figure 2. (**A**) β-Secretase (BACE1) and (**B**) acetylcholinesterase (AChE) inhibitory activities of eckol (■), dieckol (♦), and 8,8′-bieckol (▲). All assays were performed in three independent experiments. Dimethyl sulfoxide (DMSO) was used as negative controls in the BACE1 and AChE assays.

Three phlorotannins displayed high potencies as AChE inhibitors (Table 1 and Figure 2). Both dieckol and 8,8′-bieckol exhibited potent AChE inhibitory activity (IC_{50} values of 5.69 ± 0.42 and 4.59 ± 0.32 μM, respectively) and about twofold greater than the eckol (IC_{50}, 10.03 ± 0.94 μM).

To demonstrate the specificity of the targeted enzymes, all compounds were tested against tumor necrosis-converting enzyme (TACE), which is a candidate for α-secretase, and other serine proteases, including trypsin, chymotrypsin, and elastase (Table 2). With serine proteases being found in nearly all body tissues and involved in various physiological functions, including digestion, reproduction and immune response, off-target activity causing severe side effects is possible and likely if an inhibitor is not specific to BACE1 and AChE [35]. Therefore, to determine whether phlorotannins inhibited only targeted enzymes without affecting the normal pathway, all compounds were tested against TACE and serine proteases. Up to 100 μM, our tested compounds did not show significant inhibition against the above enzymes, indicating that three phlorotannins appeared to be relatively specific inhibitors of BACE1 and AChE.

Table 2. Inhibitory activities of phlorotannins against tumor necrosis-converting enzyme (TACE), trypsin, chymotrypsin, and elastase [a,b].

Compounds (μM)		TACE (α-Secretase)	Trypsin	Chymotrypsin	Elastase
Eckol	50	19.29 ± 1.52	3.59 ± 0.57	1.48 ± 0.19	6.06 ± 0.13
	100	10.60 ± 0.53	4.73 ± 0.25	2.65 ± 0.06	4.37 ± 0.27
Dieckol	50	16.90 ± 1.01	6.44 ± 0.76	4.39 ± 0.44	6.70 ± 0.85
	100	18.33 ± 0.41	−16.64 ± 1.40	1.43 ± 0.02	6.20 ± 0.14
8,8′-Bieckol	50	11.07 ± 0.53	2.98 ± 0.26	0.86 ± 0.02	5.26 ± 0.43
	100	3.57 ± 0.05	−23.05 ± 0.32	0.33 ± 0.01	7.52 ± 0.24

[a] The inhibitory activity (%) was expressed as the mean ± SD of three independent experiments. DMSO was used as a negative control in TACE and serine proteases assays. [b] Comparison of concentration level in each sample is not significantly different.

2.2. Kinetic Study of BACE1 and AChE Inhibition

As shown in Table 1 and Figure 3, Lineweaver–Burk plots for the inhibition of BACE1 by eckol, dieckol, and 8,8′-bieckol were fitted well to the noncompetitive inhibition mode in visual inspection, and the K_i values of eckol, dieckol, and 8,8′-bieckol were 31.2, 20.1, and 13.9 μM, respectively. On the other hand, our tested compounds were competitive inhibitors of AChE, where the Lineweaver– Burk plots intersected a common point on the y-axis (Table 1 and Figure 4). The K_i values of eckol, dieckol, and 8,8′-bieckol were 37.3, 12.3, and 11.4 μM, respectively, and were obtained from the Dixon plot.

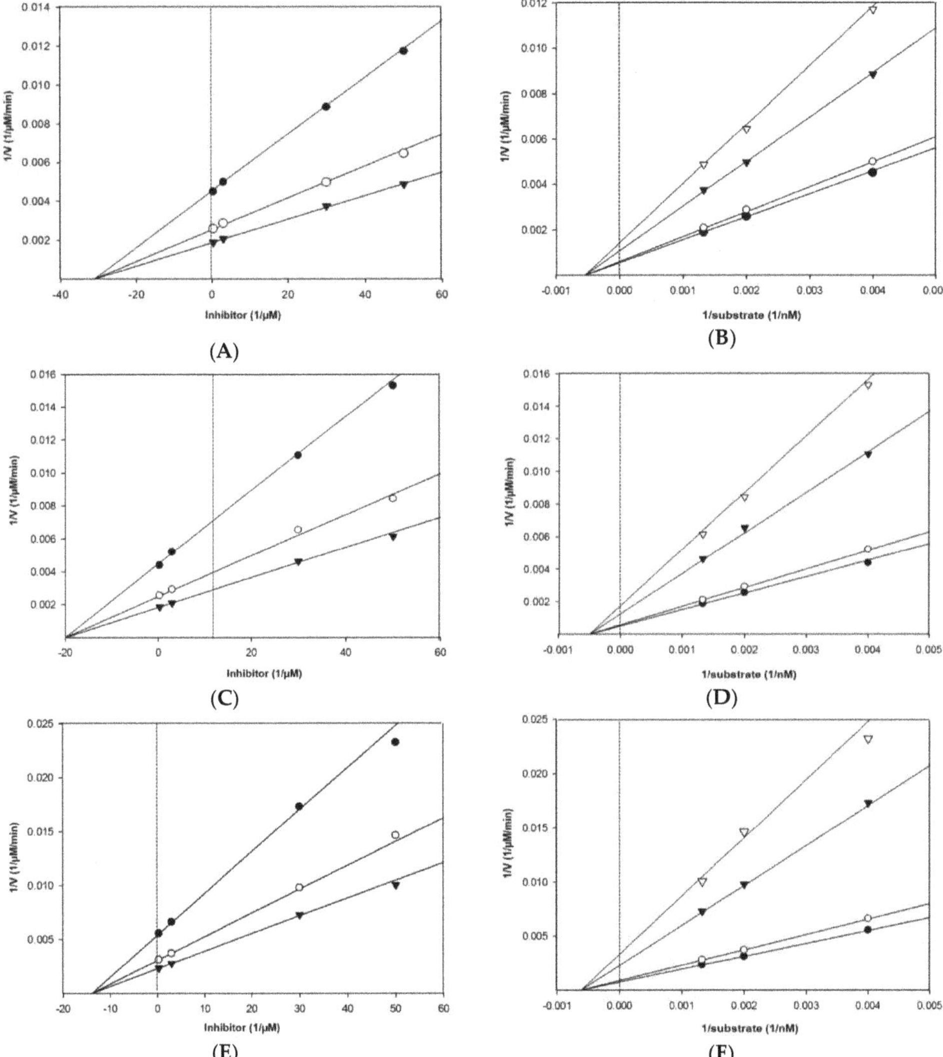

Figure 3. Dixon plot of BACE1 inhibition by (**A**) eckol, (**C**) dieckol, and (**E**) 8,8′-bieckol in the presence of different substrate concentrations: 250 nM (•), 500 nM (○), and 750 nM (▼). Lineweaver–Burk plot of BACE1 inhibition by (**B**) eckol, (**D**) dieckol, and (**F**) 8,8′-bieckol in the presence of different inhibitor concentrations: 0.3 μM (•), 3 μM (○), 30 μM (▼), and 50 μM (▽) for eckol; 0.3 μM (•), 1.5 μM (○), 3 μM (▼), and 15 μM (▽) for dieckol; 0.3 μM (•), 3 μM (○), 7.5 μM (▼), and 15 μM (▽) for 8,8′-bieckol. All assays were performed in three independent experiments.

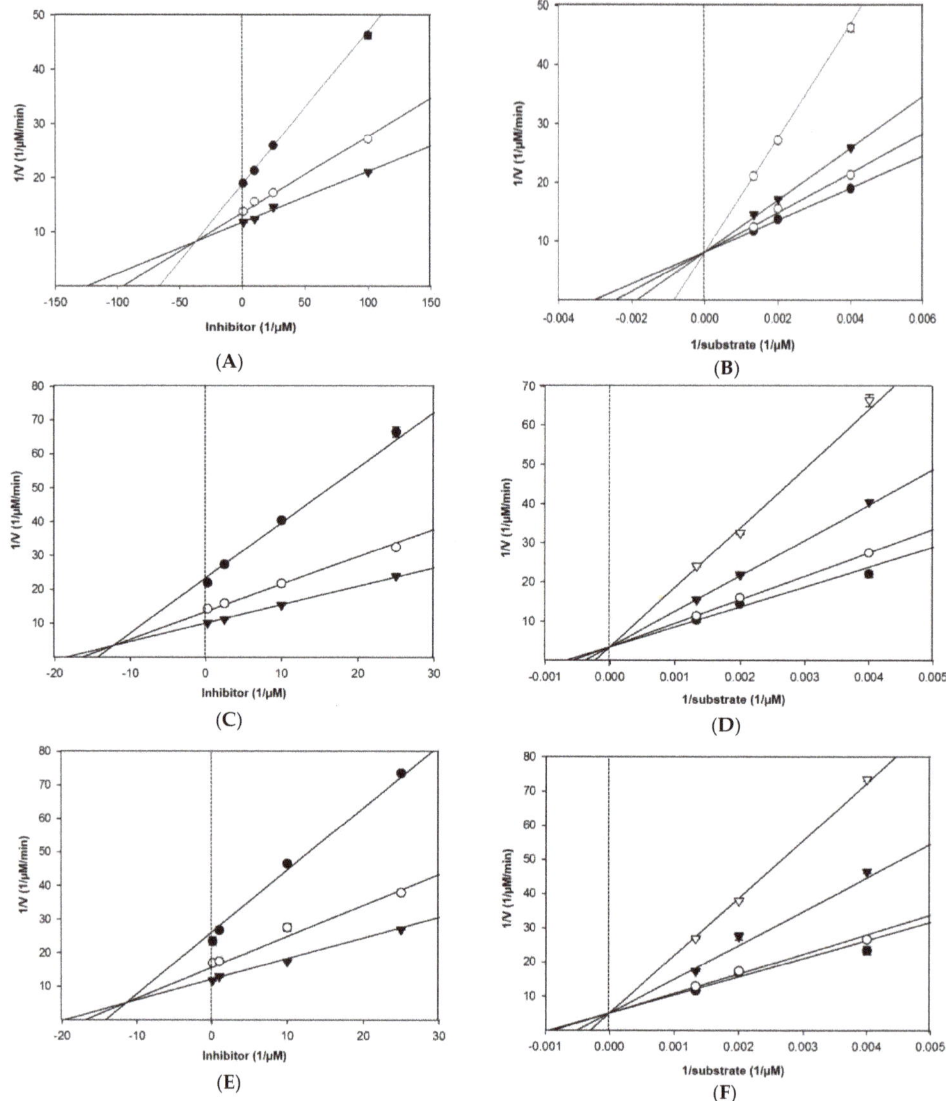

Figure 4. Dixon plot of AChE inhibition by (**A**) eckol, (**C**) dieckol, and (**E**) 8,8′-bieckol in the presence of different substrate concentrations: 250 µM (•), 500 µM (○), and 750 µM (▼). Lineweaver–Burk plot of AChE inhibition by (**B**) eckol, (**D**) dieckol, and (**F**) 8,8′-bieckol in the presence of different inhibitor concentrations: 1 µM (•), 10 µM (○), 25 µM (▼), and 100 µM (▽) for eckol; 0.1 µM (•), 10 µM (○), 25 µM (▼), and 50 µM (▽) for dieckol and 8,8′-bieckol. All assays were performed in three independent experiments.

2.3. In Silico Docking Study of the Inhibition of BACE1 and AChE by Phlorotannins

According to the in silico docking simulation results, BACE1 and phlorotannins complexes had an allosteric inhibition mode (Table 3 and Figure 5). GLY34 and SER36 of BACE1 formed two hydrogen bonds with the hydroxyl group of eckol with bonding distances of 3.277 and 3.239 Å, respectively. In the dieckol–BACE1 complex, TRP76, THR232, and LYS321 participated in three hydrogen bonds (bonding distance: 2.960, 3.149, and 3.488 Å, respectively). 8,8′-Bieckol–BACE1 complex had four hydrogen bonding interactions with residues LYS107, GLY230, THR231, and SER325 (bonding distance: 3.120, 2.773, 3.098, and 2.887 Å, respectively). In addition, the lowest binding energy of the three tested compounds were negative values: -8.8 kcal/mol for eckol, -10.1 kcal/mol for dieckol, and -9.0 kcal/mol for 8,8′-bieckol.

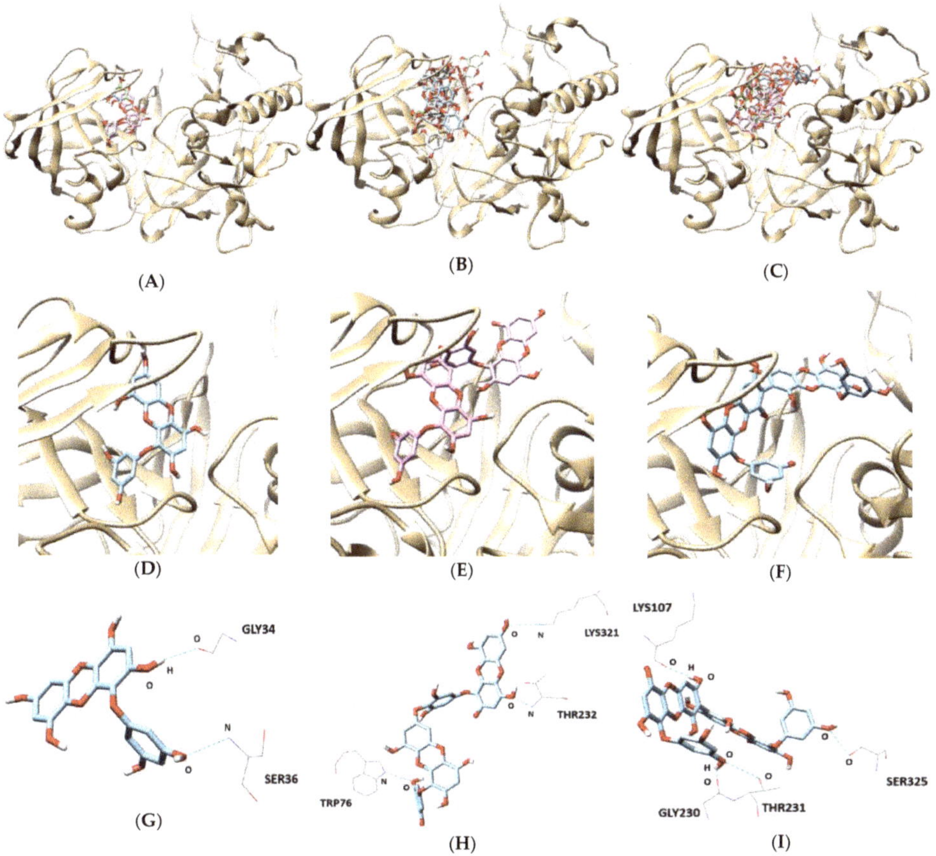

Figure 5. In silico docking simulation of BACE1 inhibition by (**A**) eckol, (**B**) dieckol, and (**C**) 8,8′-bieckol. View of the binding site magnified from (**D**) eckol, (**E**) dieckol, and (**F**) 8,8′-bieckol. Hydrogen interaction diagram of (**G**) eckol, (**H**) dieckol, and (**I**) 8,8′-bieckol.

Table 3. Molecular interactions of BACE with eckol, dieckol, and 8,8′-bieckol.

Ligand	Lowest Energy (Kcal/mol)	No. of H-Bonds	H-Bonds Interaction Residues	Bond Distance (Å)
Eckol	−8.8	2	GLY34	3.277
			SER36	3.239
Dieckol	−10.1	3	TRP76	2.960
			THR232	3.149
			LYS321	3.488
8,8′-Bieckol	−9.0	4	LYS107	3.120
			GLY230	2.773
			THR231	3.098
			SER325	2.887

As indicated in Table 4 and Figure 6, the docking results for eckol, dieckol, and 8,8′-bieckol indicated negative binding energies of −8.8, −9.5, and −9.2 kcal/mol, respectively. Hydrogen bonding interactions between eckol and THR83, TRP86, TYR124, and SER125 of AChE were observed by five hydrogen bonds (bonding distance of 2.855, 2.712, 3.134, 2.883, and 3.313 Å, respectively). Dieckol was bound at the ASN233, THR238, ARG296, and HIS405 of AChE, linked by four hydrogen bonds with bonding distances of 3.399, 2.837, 3.344, and 3.181 Å, respectively, while 8,8′-bieckol had one hydrogen bond with the ARG296 residue of AChE (bonding distance: 3.151 Å).

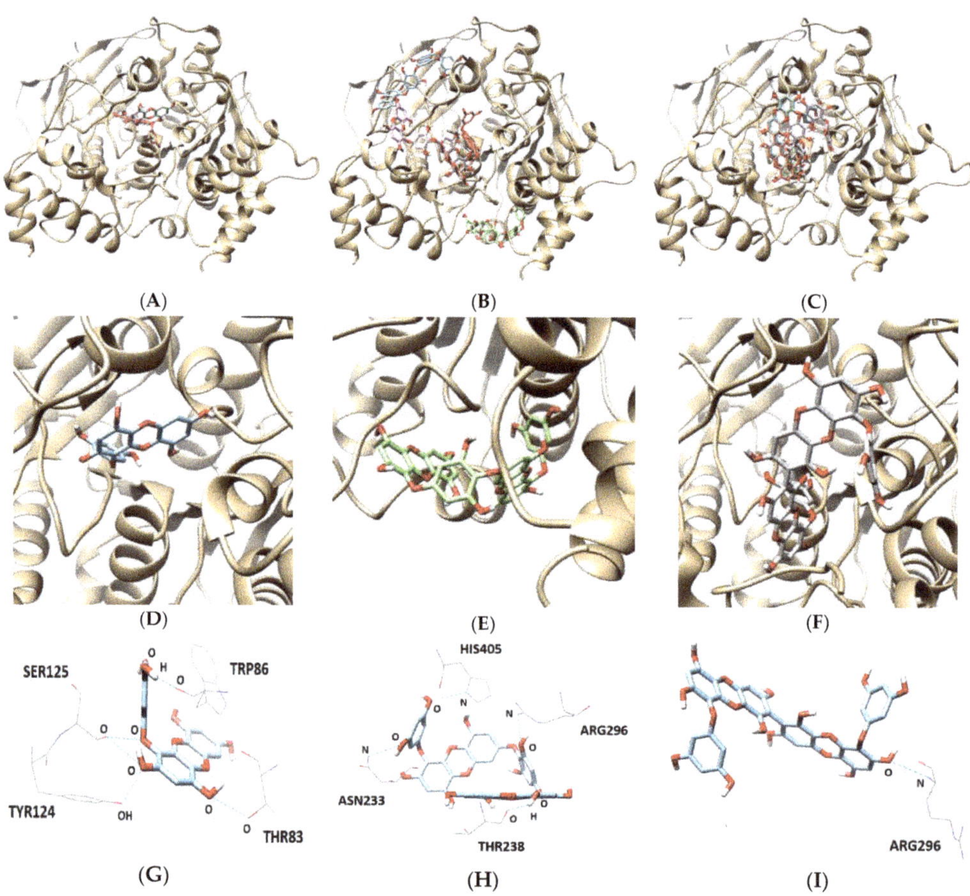

Figure 6. In silico docking simulation of AChE inhibition by (**A**) eckol, (**B**) dieckol, and (**C**) 8,8′-bieckol. View of the binding site magnified from (**D**) eckol, (**E**) dieckol, and (**F**) 8,8′-bieckol. Hydrogen interaction diagram of (**G**) eckol, (**H**) dieckol, and (**I**) 8,8′-bieckol.

Table 4. Molecular interactions of AChE with eckol, dieckol, and 8,8′-bieckol.

Compounds	Lowest Energy (Kcal/mol)	No. of H-Bonds	H-Bonds Interaction Residues	Bond Distance (Å)
Eckol	−8.8	5	THR83 TRP86 TYR124 SER125	2.855 2.712 3.134 2.883 & 3.313
Dieckol	−9.5	4	ASN233 THR238 ARG296 HIS405	3.399 2.837 3.344 3.181
8,8′-Bieckol	−9.2	1	ARG296	3.151

3. Discussion

With respect to the development of anti-AD agents, enzyme inhibition is one of the most promising potential therapeutic strategies. Since BACE1 is the initiating and rate-limiting enzyme in Aβ formation, it is considered as a key target for lowering cerebral Aβ levels [7–9]. Additionally, AChE plays a critical role in cholinergic neurotransmission and participates in non-cholinergic mechanisms such as accelerating Aβ fibril formation through conformational change of Aβ and increasing Aβ toxicity by Aβ–AChE complexes [14–16]. Thus, multi-enzyme target inhibition against BACE1 and AChE may provide a promising therapeutic approach for AD. In search of a candidate for AD prevention and/or treatment, numerous researchers over the past few decades have focused on discovering natural enzyme inhibitors. Several natural inhibitors of BACE1 and AChE such as coumarins, citrus flavanones, triterpenoids, and alkaloids have been reported [36–39]. However, efforts to explore bioactive constituents from marine organisms against BACE1 and AChE have been relatively limited.

In the current study, three effective phlorotannins—eckol, dieckol, and 8,8′-bieckol—were studied for their inhibitory properties on BACE1 and AChE. These compounds exhibited powerful inhibitory activities on BACE1 with IC_{50} values at a range of 1.6–7.7 μM. Several terrestrial plant-derived BACE1 inhibitors, including hespretin, naringenin, and hesperidin, were from citrus fruits with IC_{50} values ranging from 16.9–30.3 μM. Alkaloids (neferine, liensinine, and vitexin) in *Nelumbo nucifera* (IC_{50} in the 6.4–28.5 μM range) have been proven to be efficient BACE1 inhibitors. Umbelliferone, isoscopoletin, 7-methoxy coumarin, esculetin, and daphnetin from *Angelica decursiva* with IC_{50} ranging from 7.7–172.3 μM were identified as BACE1 inhibitors. Compared with those plant-derived BACE1 inhibitors, our compounds demonstrated predominantly inhibitory properties against BACE1 [36,37,39].

Interestingly, the difference in inhibitory properties among phlorotannins is related to the number of hydroxyl groups present. In our new findings, it was shown that 8,8′-bieckol containing 11 OH groups had the highest inhibitory efficacy against BACE1 when compared to dieckol (10 OH groups) and eckol (6 OH groups). When phlorotannins from *Eisenia bicyclis*, one of the brown algae, were investigated for their BACE1 inhibitory effects, the result that dieckol was stronger than eckol was similar to that of our present study [40]. Consistent with our result, Ahn and colleagues have reported that the inhibitory effect on HIV-reverse transcriptase of 8,8′-bieckol containing a biaryl linkage was tenfold higher than that of 8,4‴-dieckol with a diphenyl ether linkage [41]. This observation indicated that the steric hindrance of the hydroxyl and aryl groups near the biaryl linkage of 8,8′-bieckol noticeably enhanced its inhibitory potency.

Among three phlorotannins, 8,8′-bieckol showed the most potent inhibitory activity against AChE. Similar results regarding the correlation between the molecular size of phlorotannins and enzyme inhibitory efficacy was revealed in a previous study. 8,8′-bieckol showed more potent activity against hyaluronidase, with an IC_{50} of 40 μM, than dieckol and eckol (IC_{50}, 120 and >800 μM, respectively) [42]. In addition, the present study first demonstrated the specific molecular docking interaction as well as biological properties of eckol, dieckol, and 8,8′-bieckol against AChE.

The BACE1 inhibition kinetics indicated that the tested compounds act as non-competitive inhibitors, which means that these compounds can bind either another regulatory site or to the subsite of BACE1. The inhibition level is dependent on the concentration of the inhibitor but is not reduced by increasing concentrations of substrate. Because of this, V_{max} is reduced, but K_m is unaffected. In BACE1 inhibitory activity, phlorotannins decreased the V_{max} values without affecting the affinity of BACE1 toward the K_m, which demonstrated that these compounds exhibited non-competitive inhibition against BACE1. However, AChE kinetics results exhibited that our tested compounds are competitive inhibitors with unchanged V_{max} and increased K_m. In other words, these compounds interacted directly with the catalytic site of AChE instead of with other allosteric pockets.

In silico docking analysis is a valuable drug discovery tool and can be used to discover prospective, biologically active molecules from natural product databases. The results of the molecular docking score were provided to evaluate the capacity of different protein–ligand complex interactions and to compare the biological activities and the inhibition mode. In the BACE1 docking simulation, multiple hydrogen interactions were observed in the BACE1–phlorotannins complexes. Eckol interacted with both GLY34 and SER36 of BACE1, and dieckol bounded to TRP76, THR232, and LYS321. In addition, 8,8′-bieckol formed four hydrogen bonds with BACE1 residues, including LYS107, GLY230, THR231, and SER325. These docking results showed that hydrogen bonds between phlorotannins and allosteric residues of BACE1 play an important role in enzyme inhibition.

AChE docking analysis provides insight into the mechanism underlying active site binding interaction. The hydroxyl group of eckol formed five hydrogen bonds with THR83, TRP86, TYR124, and SER125 of AChE. In particular, the choline-binding site residue (TRP86) of AChE was involved in hydrogen bond interaction with eckol. Dieckol showed four hydrogen-bond interactions with ASN233, THR238, ARG296, and HIS405, whereas 8,8′-bieckol made one hydrogen bond with ARG296 located in the active site of AChE. These docking results from the in silico study were in agreement with our in vitro experimental data.

To date, few studies have investigated the neuroprotective property of phlorotannins. Our previous study demonstrated that phlorotannins ameliorated $A\beta_{25-35}$ toxicity through the regulation of the apoptotic signal and the NF-kB/MAPKs pathway [34]. Eckol and dieckol suppressed H_2O_2-induced oxidative stress in murine hippocampus neuronal cells [43]. Moreover, it has been reported in an in vivo study that oral administration of dieckol (10 mg/kg) improved cognitive ability in ethanol-induced memory impairment mice [44].

Nagayama and coworkers demonstrated no significant toxic effects in the oral administration of up to 1,500 mg/kg of phlorotannins for 14 days in male and female Institute of Cancer Research (ICR) mice [45]. In a human study, *E. Cava* extract was shown to be safe for use in food supplements at a maximum daily intake level of 263 mg/day for adults [46]. Collectively, phlorotannins are toxicologically very safe, explaining their traditional and present consumption as foods and medicinal products.

Bioavailability parameters such as biotransformation and conjugation during absorption from the GI tract are principle factors influencing in vivo biological activity. Lipinski's rule of five is a widespread strategy to define bioavailability predictions of drug molecules. According to this predictive model, a compound needs to exhibit optimum GI absorption with a molecular weight of < 500 Da, no more than five hydrogen bond donors, no more than ten hydrogen bond acceptors, and a calculated partition coefficient (LogP) that is no more than five [47]. Fortunately, eckol meets Lipinski's requirements for acceptable oral bioavailability, while dieckol and 8,8′-bieckol have limitations on bioavailability [48]. However, the compounds absorbed by specific transporters are an exception to this rule, and a recent study demonstrated that dieckol successfully penetrated into the brain via crossing the blood–brain barrier (BBB), suggesting that the compound may be transported through an unknown mechanism [49]. A study of the permeability of eckol and 8,8′-bieckol was limited, but it is likely that similar results might also be predictable as that of dieckol. Overall, our marine compounds from *E. Cava* are safe, potent, and selective natural dual inhibitors against BACE1 and AChE that can be used for the multi-target, directed agents of AD.

4. Materials and Methods

4.1. General

Fluorescence and optical density were measured by Bio-TEK ELISA fluorescence reader FLx 800 and Bio-TEK ELx 808, respectively (Winooski, VT, USA). Eckol (>95%), dieckol (>95%), and 8,8′-bieckol (>95%), were bought from National Development Institute of Korean Medicine (Gyeongsangbuk-do, Korea). The BACE1 assay kit was purchased from Invitrogen (Pan Vera, Madison, WI, USA). TACE and substrate were bought from R&D Systems (Minneapolis, MN, USA). AChE from *Electrophorus electricus* (electric eel), 5,5′-dithiobis-(2 nitrobenzoic acid) (DTNB), resveratrol, galantamine, trypsin, chymotrypsin, elastase, and their substrates, including N-benzoyl-L-Arg-pNA, N-benzoyl-L-Tyr-pNA, and N-succinyl-Ala-Ala-Ala-p NA, were from Sigma-Aldrich (St. Louis, MO, USA).

4.2. Enzyme inhibition Studies

Fluorometric assays with a recombinant human BACE1 or TACE were conducted according to manufacturer instructions. Briefly, reaction mixtures containing human recombinant BACE1 (1.0 U/mL), the substrate (75 μM in 50 mM ammonium bicarbonate), and phlorotannins dissolved in an assay buffer (50 mM sodium acetate, pH 4.5) were incubated in darkness for 60 min at 25 °C in well plates. The increase in fluorescence intensity produced by substrate hydrolysis was observed on a fluorescence microplate reader with excitation and emission wavelengths of 545 and 590 nm, respectively. The inhibition ratio was obtained using the following equation:

$$\text{Inhibition (\%)} = [1 - (S - S_0)/(C - C_0)] \times 100$$

where *C* was the fluorescence of control (enzyme, assay buffer, and substrate) after 60 min of incubation, C_0 was the fluorescence of control at time 0, *S* was the fluorescence of tested samples (enzyme, sample solution, and substrate) after 60 min of incubation, and S_0 was the fluorescence of the tested samples at time 0.

A human recombinant TACE (0.1 ppm in 25 mM Tris buffer), the substrate (APP peptide YEVHHQKLV using EDANS/DABCYL), and phlorotannins were dissolved in an assay buffer, which were then combined and incubated for 60 min in the dark at 25 °C. The increase in fluorescence intensity produced by substrate hydrolysis was observed on a fluorescence microplate reader with excitation and emission wavelengths of 320 and 405 nm, respectively.

The colorimetric assays, including AChE, trypsin, chymotrypsin, and elastase were assayed according to previously described methods [36]. The hydrolysis of AChE was monitored according to the formation of yellow 5-thio-2-nitrobenzoate anions at 405 nm for 15 min, which were produced by the reaction of DTNB with thiocholine released from ACh. All reactions were performed in 96-well plates in triplicate and recorded using a microplate spectrophotometer.

N-benzoyl-L-Arg-*p*NA, *N*-benzoyl-L-Tyr-*p*NA, and *N*-succinyl-Ala-Ala-Ala-*p*NA were used as substrates to assay the inhibition of trypsin, chymotrypsin, and elastase, respectively. Enzyme, Tris-HCl buffer (0.05 M, in 0.02 M $CaCl_2$, pH 8.2), and phlorotannins were incubated for 10 min at 25 °C; then, substrate was added for 30 min at 37 °C. The absorbance was recorded at 410 nm. The inhibition ratio was obtained using the following equation:

$$\text{Inhibition (\%)} = \{[1 - (\text{A} - \text{B})]/\text{control}\} \times 100$$

where A was the absorbance of the control (enzyme, assay buffer, and substrate) after 60 min of incubation, and B was the absorbance of tested sample (assay buffer and sample solution) after 60 min of incubation.

4.3. Enzyme Kinetic Study

To evaluate the kinetic mechanisms of phlorotannins towards BACE1 and AChE, Dixon and Lineweaver–Burk plots were conducted by various concentrations of substrate and inhibitors. Kinetic parameters such as K_i, V_{max}, and K_m values were calculated by Sigma Plot 12.3 (Systat Software, Inc., San Jose, CA, USA)

4.4. Molecular Docking Study

X-ray crystal structures of human BACE1 (PDB code: 2WJO) and AChE (PDB code: 4PQE) were retrieved from the Protein Data Bank (PDB, http://www.rcsb.org/). Three-dimensional (3D) structures of eckol, dieckol, and 8,8′-bieckol were obtained from PubChem with compound identification number (CID) of 145937, 3008868, and 3008867, respectively. The Autodock Vina software version 1.1.2 (The Scripps Research Institute, La Jolla, CA, USA,) was used to conduct molecular docking analysis. The dimensions of the grid were 30 × 30 × 30 Å, the cluster radius was 1 Å, and the Cα coordinates in each selected backbone binding residue of the protein receptor was used for the center of docking space. Other options for docking simulations were used as defaults. The atomic coordinates of the ligands were drawn and displayed using Marvin sketch (5.11.4, 2012, ChemAxon, One Broadway Cambridge, MA, USA).

4.5. Statistical Analysis

All results were presented as the mean ± SD of three independent experiments. Statistical significance was assessed by Duncan′s multiple range tests using Statistical Analysis System (SAS) version 9.3 (SAS Institute, Cary, NC, USA).

5. Conclusions

The integration of enzyme activity, kinetics, and in silico docking studies provided principle insights into the molecular basis underlying ligand binding affinity and BACE1 and AChE inhibition. Accordingly, these results suggested that phlorotannins from *E. cava*, especially 8,8′-bieckol, have noteworthy potential for the possible development as treatments and/or preventative agents against AD.

Author Contributions: Data curation, writing-original draft, J.L.; Supervision, review and editing, M.J.

Funding: This research was funded by Dong-A University.

Conflicts of Interest: All authors have no conflict of interest to declare.

References

1. Selkoe, D. Alzheimer's disease results from the cerebral accumulation and cytotoxicity of amyloid beta-protein. *J. Alzheimers Dis.* **2001**, *3*, 75–80. [CrossRef] [PubMed]
2. Götz, J.; Schild, A.; Hoerndli, F.; Pennanen, L. Amyloid-induced neurofibrillary tangle formation in Alzheimer's disease: Insight from transgenic mouse and tissue-culture models. *Int. J. Dev. Neurosci.* **2004**, *7*, 453–465.
3. Lin, G.; Koelsch, S.; Dashti, A. Human aspartic protease memapsin 2 cleaves the β-secretase site of Aβ-amyloid precursor protein. *Proc. Natl. Acad. Sci. USA* **2000**, *97*, 1456–1460. [CrossRef] [PubMed]
4. Bittner, T.; Fuhrmann, M.; Burgold, S.; Jung, C.; Volbracht, C.; Steiner, H.; Mitteregger, G.; Kretzschmar, H.; Haass, C.; Herms, J. Gamma-secretase inhibition reduces spine density *in vivo via* an amyloid precursor protein-dependent pathway. *J. Neurosci.* **2009**, *29*, 10405–10409. [CrossRef] [PubMed]
5. De Strooper, B. Lessons from a failed γ-secretase Alzheimer trial. *Cell* **2014**, *159*, 721–726. [CrossRef] [PubMed]
6. Peters, F.; Salihoglu, H.; Rodrigues, E.; Herzog, E.; Blume, T.; Filser, S.; Dorostkar, M.; Shimshek, D.; Brose, N.; Neumann, U.; et al. BACE1 inhibition more effectively suppresses initiation than progression of β-amyloid pathology. *Acta Neuropathol.* **2018**, *135*, 695–710. [CrossRef] [PubMed]

7. Neumann, U.; Rueeger, H.; Machauer, R.; Veenstra, S.; Lueoend, R.; Tintelnot-Blomley, M.; Laue, G.; Beltz, K.; Vogg, B.; Schmid, P.; et al. A novel BACE inhibitor NB-360 shows a superior pharmacological profile and robust reduction of amyloid-β and neuroinflammation in APP transgenic mice. *Mol. Neurodegenr.* **2015**, *10*, 44–58. [CrossRef]
8. Thakker, D.; Sankaranarayanan, S.; Weatherspoon, M.; Harrison, J.; Pierdomenico, M.; Heisel, J.; Pierdomenico, M.; Heisel, J.; Thompson, L.; Haskell, R.; et al. Centrally delivered BACE1 inhibitor activates microglia and reverses amyloid pathology and cognitive deficit in aged Tg2576 mice. *J. Neurosci.* **2015**, *35*, 6931–6936. [CrossRef]
9. Kennedy, M.; Stamford, A.; Chen, X.; Cox, K.; Cumming, J.; Dockendorf, M.; Egan, M.; Ereshefsky, L.; Hodgson, R.; Hyde, L.; et al. The BACE1 inhibitor verubecestat (MK-8931) reduces CNS β-amyloid in animal models and in Alzheimer's disease patients. *Sci. Transl. Med.* **2016**, *8*, 363ra150. [CrossRef]
10. Cheret, C.; Willem, M.; Fricker, F.; Wende, H.; Wulf-Goldenberg, A.; Tahirovic, S.; Nave, K.; Saftig, P.; Haass, C.; Garratt, A.; et al. Bace1 and Neuregulin-1 cooperate to control formation and maintenance of muscle spindles. *EMBO J.* **2013**, *32*, 2015–2028. [CrossRef]
11. Filser, S.; Ovsepian, S.; Masana, M.; Blazquez-Llorca, L.; Brandt Elvang, A.; Volbracht, C.; Müller, M.; Jung, C.; Herms, J. Pharmacological inhibition of BACE1 impairs synaptic plasticity and cognitive functions. *Biol. Psychiatry.* **2015**, *77*, 729–739. [CrossRef] [PubMed]
12. Zhu, K.; Xiang, X.; Filser, S.; Marinković, P.; Dorostkar, M.; Crux, S.; Neumann, U.; Shimshek, D.; Rammes, G.; Haass, C.; et al. Beta-site amyloid precursor protein cleaving enzyme 1 inhibition impairs synaptic plasticity via seizure protein 6. *Biol. Psychiatry* **2018**, *83*, 428–437. [CrossRef] [PubMed]
13. Russo, P.; Kisialiou, A.; Lamonaca, P.; Moroni, R.; Prinzi, G.; Fini, M. New drugs from marine organisms in alzheimer's disease. *Mar. Drugs* **2016**, *14*, 5. [CrossRef] [PubMed]
14. Rees, T.; Hammond, P.; Soreq, H.; Younkin, S.; Brimijoin, S. Acetylcholinesterase promotes beta-amyloid plaques in cerebral cortex. *Neurobiol. Aging* **2003**, *24*, 777–787. [CrossRef]
15. Inestrosa, N.; Dinamarca, M.; Alvarez, A. Amyloid–cholinesterase interactions. *FEBS J.* **2008**, *275*, 625–632. [CrossRef] [PubMed]
16. Dinamarca, M.; Sagal, J.; Quintanilla, R.; Godoy, J.; Arrázola, M.; Inestrosa, N. Amyloid-beta-Acetylcholinesterase complexes potentiate neurodegenerative changes induced by the Abeta peptide. Implications for the pathogenesis of Alzheimer's disease. *Mol. Neurodegener.* **2010**, *5*, 4–19. [CrossRef] [PubMed]
17. Prado-Prado, F.; García, I. Review of theoretical studies for prediction of neurodegenerative inhibitors. *Mini Rev. Med. Chem.* **2012**, *12*, 452–466. [CrossRef]
18. Folch, J.; Petrov, D.; Ettcheto, M.; Abad, S.; Sánchez-López, E.; García, M.; Olloquequi, J.; Beas-Zarate, C.; Auladell, C.; Camins, A. Current research therapeutic strategies for Alzheimer's disease treatment. *Neural Plast.* **2016**, *2016*, 8501693. [CrossRef]
19. Kang, I.; Jeon, Y.; Yin, X.; Nam, J.; You, S.; Hong, M.; Jang, B.; Kim, M. Butanol extract of *Ecklonia cava* prevents production and aggregation of beta-amyloid, and reduces beta-amyloid mediated neuronal death. *Food Chem. Toxicol.* **2011**, *49*, 2252–2259. [CrossRef]
20. Velmurugan, B.; Rathinasamy, B.; Lohanathan, B.; Thiyagarajan, V.; Weng, C. Neuroprotective Role of Phytochemicals. *Molecules* **2018**, *23*, 2485. [CrossRef]
21. Kang, M.-C.; Kang, N.; Ko, S.-C.; Kim, Y.-B.; Jeon, Y.-J. Anti-obesity effects of seaweeds of Jeju Island on the differentiation of 3T3-L1 preadipocytes and obese mice fed a high-fat diet. *Food Chem. Toxicol.* **2016**, *90* (Suppl. C), 36–44. [CrossRef] [PubMed]
22. Lee, S.-H.; Ko, S.-C.; Kang, M.-C.; Lee, D.; Jeon, Y.-J. Octaphlorethol A, a marine algae product, exhibits antidiabetic effects in type 2 diabetic mice by activating AMP-activated protein kinase and upregulating the expression of glucose transporter 4. *Food Chem. Toxicol.* **2016**, *91* (Suppl. C), 58–64. [CrossRef] [PubMed]
23. Sanjeewa, K.; Fernando, I.; Samarakoon, K.; Lakmal, H.; Kim, E.-A.; Kwon, O.; Dilshara, M.; Lee, J.-B.; Jeon, Y.-J. Anti-inflammatory and anti-cancer activities of sterol rich fraction of cultured marine microalga *Nannochloropsis oculata*. *Algae* **2016**, *31*, 277–287. [CrossRef]
24. Ko, S.-C.; Jung, W.-K.; Lee, S.-H.; Lee, D.; Jeon, Y.-J. Antihypertensive effect of an enzymatic hydrolysate from Styela clava flesh tissue in type 2 diabetic patients with hypertension. *Nutr. Res. Pract.* **2017**, *11*, 396–401. [CrossRef] [PubMed]

25. Wang, L.; Lee, W.; Oh, J.; Cui, Y.; Ryu, B.; Jeon, Y.-J. Protective effect of sulfated polysaccharides from celluclast-assisted extract of Hizikia fusiforme against ultraviolet-B-induced skin damage by regulating NF-κB, AP-1, and MAPKs signaling pathways in vitro in human dermal fibroblasts. *Mar. Drugs* **2018**, *16*, 239. [CrossRef] [PubMed]
26. Wijesekara, I.; Yoon, N.; Kim, S. Phlorotannins from *Ecklonia cava* (Phaeophyceae): Biological activities and potential health benefits. *Biofactors* **2010**, *36*, 408–414. [CrossRef]
27. Shibata, T.; Kawaguchi, S.; Hama, Y.; Inagaki, M.; Yamaguchi, K.; Nakamura, T. Local and chemical distribution of phlorotannins in brown algae. *J. Appl. Phycol.* **2004**, *16*, 291–296. [CrossRef]
28. Ahn, G.; Kim, K.; Cha, S.; Song, C.; Lee, J.; Heo, M.; Yeo, I.; Lee, N.; Jee, Y.; Kim, J.; et al. Antioxidant activities of phlorotannins purified from *Ecklonia cava* on free radical scavenging using ESR and H_2O_2-mediated DNA damage. *Eur. Food Res. Technol.* **2007**, *226*, 71–79. [CrossRef]
29. Lee, S.; Li, Y.; Karadeniz, F.; Kim, M.; Kim, S. a–Glycosidase and a–amylase inhibitory activities of phloroglucinal derivatives from edible marine brown alga, *Ecklonia cava*. *J. Sci. Food Agric.* **2009**, *89*, 1552–1558. [CrossRef]
30. Athukorala, Y.; Jeon, Y.J. Screening for Angiotensin 1-converting enzyme inhibitory activity of *Ecklonia cava*. *J. Food Sci. Nutr.* **2005**, *10*, 134–139. [CrossRef]
31. Artan, M.; Li, Y.; Karadeniz, F.; Lee, S.; Kim, M.; Kim, S. Anti-HIV-1 activity of phloroglucinol derivative, 6,6′ -bieckol, from *Ecklonia cava*. *Bioorg. Med. Chem.* **2008**, *16*, 7921–7926. [CrossRef] [PubMed]
32. Zhang, R.; Kang, K.; Piao, M.; Ko, D.; Wang, Z.; Lee, I.; Kim, B.; Jeong, I.; Shin, T.; Park, J.; et al. Eckol protects V79-4 lung fibroblast cells against t-ray radiation-induced apoptosis via the scavenging of reactive oxygen species and inhibiting of the c-Jun NH2-terminal kinase pathway. *Eur. J. Pharmacol.* **2008**, *591*, 114–123. [CrossRef] [PubMed]
33. Kang, I.; Jang, B.; In, S.; Choi, B.; Kim, M.; Kim, M. Phlorotannin-rich Ecklonia cava reduces the production of beta-amyloid by modulating alpha- and gamma-secretase expression and activity. *Neurotoxicology* **2013**, *34*, 16–24. [CrossRef] [PubMed]
34. Lee, S.; Youn, K.; Kim, D.; Ahn, M.-R.; Yoon, E.; Kim, O.-Y.; Jun, M. Anti-Neuroinflammatory Property of Phlorotannins from *Ecklonia cava* on $A\beta_{25\text{-}35}$-Induced Damage in PC12 Cells. *Mar. Durgs* **2019**, *17*, 7. [CrossRef] [PubMed]
35. Hedstrom, L. Serine protease mechanism and specificity. *Chem. Rev.* **2002**, *102*, 4501–4524. [CrossRef] [PubMed]
36. Ali, M.; Jannat, S.; Jung, H.; Choi, R.; Roy, A.; Choi, J. Anti-Alzheimer's disease potential of coumarins from *Angelica decursiva* and *Artemisia capillaris* and structure-activity analysis. *Asian Pac. J. Trop. Med.* **2016**, *9*, 103–111. [CrossRef]
37. Lee, S.; Youn, K.; Lim, G.; Lee, J.; Jun, M. In Silico Docking and *In Vitro* Approaches towards BACE1 and Cholinesterases Inhibitory Effect of Citrus Flavanones. *Molecules* **2018**, *23*, 1509. [CrossRef]
38. Nguyen, V.; Zhao, B.; Seong, S.; Kim, J.; Woo, M.; Choi, J.; Min, B. Inhibitory effects of serratene-type triterpenoids from Lycopodium complanatum on cholinesterases and β-secretase 1. *Chem. Biol. Interact.* **2017**, *274*, 150–157. [CrossRef]
39. Jung, H.; Karki, S.; Kim, J.; Choi, J. BACE1 and cholinesterase inhibitory activities of *Nelumbo nucifera* embryos. *Arch. Pharm. Res.* **2015**, *38*, 1178–1187. [CrossRef]
40. Jung, H.; Oh, S.; Choi, J. Molecular docking studies of phlorotannins from *Eisenia bicyclis* with BACE1 inhibitory activity. *Bioorg. Med. Chem. Lett.* **2010**, *20*, 3211–3215. [CrossRef]
41. Ahn, M.-J.; Yoon, K.-D.; Min, S.-Y.; Lee, J.; Kim, J.; Kim, T.; Kim, S.; Kim, N.-K.; Huh, H.; Kim, J. Inhibition of HIV-1 Reverse Transcriptase and Protease by Phlorotannins from the Brown Alga *Ecklonia cava*. *Biol. Pharm. Bull.* **2004**, *27*, 544–547. [CrossRef] [PubMed]
42. Shibata, T.; Fujimoto, K.; Nagayama, K.; Yamaguchi, K.; Nakamura, T. Inhibitory activity of brown algal phlorotannins against hyaluronidase. *Int. J. Food Sci. Technol.* **2002**, *37*, 703–709. [CrossRef]
43. Kang, S.; Cha, S.; Ko, J.; Kang, M.; Kim, D.; Heo, S.; Kim, J.; Heu, M.; Kim, Y.; Jung, W.; et al. Neuroprotective effects of phlorotannins isolated from a brown alga, *Ecklonia cava*, against H_2O_2-induced oxidative stress in murine hippocampal HT22 cells. *Environ. Toxicol. Pharmacol.* **2012**, *34*, 96–105. [CrossRef] [PubMed]
44. Myung, C.; Shin, H.; Bao, H.; Yeo, S.; Lee, B.; Kang, J. Improvement of memory by dieckol and phlorofucofuroeckol in ethanol-treated mice: Possible involvement of the inhibition of acetylcholinesterase. *Arch. Pharm. Res.* **2005**, *28*, 691–698. [CrossRef] [PubMed]

45. Nagayama, K.; Iwamura, Y.; Shibata, T.; Hirayama, I.; Nakamura, T. Bactericidal activity of phlorotannins from the brown alga *Ecklonia kurome*. *J. Antimicrob. Chemoth.* **2002**, *50*, 889–893. [CrossRef]
46. Turck, D.; Bresson, J.-L.; Burlingame, B.; Dean, T.; Fairweather-Tait, S.; Heinonen, M.; Hirsch-Ernst, K.; Mangelsdorf, I.; McArdle, H.; Naska, A.; et al. Safety of *Ecklonia cava* phlorotannins as a novel food pursuant to Regulation (EC) No 258/97. *EFSA J.* **2017**, *15*, 5003–5018.
47. Abourashed, E. Bioavailability of Plant-Derived Antioxidants. *Antioxidants* **2013**, *2*, 309–325. [CrossRef]
48. Rengasamy, K.; Kulkarni, M.; Stirk, W.; Staden, J. Advances in algal drug research with emphasis on enzyme inhibitors. *Biotechnol. Adv.* **2014**, *32*, 1364–1381. [CrossRef]
49. Kwak, J.; Yang, Z.; Yoon, B.; He, Y.; Uhm, S.; Shin, H.; Lee, B.; Yoo, Y.; Lee, K.; Han, S.; et al. Blood-brain barrier-permeable fluorone-labeled dieckols acting as neuronal ER stress signaling inhibitors. *Biomaterials* **2015**, *61*, 52–60. [CrossRef]

Article

Effect of Pentacyclic Guanidine Alkaloids from the Sponge *Monanchora pulchra* on Activity of α-Glycosidases from Marine Bacteria

Irina Bakunina [1,*], Galina Likhatskaya [1], Lubov Slepchenko [1,2], Larissa Balabanova [1,2], Liudmila Tekutyeva [2], Oksana Son [2], Larisa Shubina [1] and Tatyana Makarieva [1]

1 G.B. Elyakov Pacific Institute of Bioorganic Chemistry, Far Eastern Branch, Russian Academy of Sciences, Vladivostok 690022, Russia; galin56@mail.ru (G.L.); lubov99d@mail.ru (L.S.); lbalabanova@mail.ru (L.B.); shubina@piboc.dvo.ru (L.S.); makarieva@piboc.dvo.ru (T.M.)

2 Far Eastern Federal University, Russky Island, Vladivostok 690022, Russia; tekuteva.la@dvfu.ru (L.T.); oksana_son@bk.ru (O.S.)

* Correspondence: bakun@list.ru; Tel.: +7-432-231-0705-3; Fax: +7-432-231-0705-7

Received: 12 November 2018; Accepted: 25 December 2018; Published: 1 January 2019

Abstract: The effect of monanchomycalin B, monanhocicidin A, and normonanhocidin A isolated from the Northwest Pacific sample of the sponge *Monanchora pulchra* was investigated on the activity of α-galactosidase from the marine γ-proteobacterium *Pseudoalteromonas* sp. KMM 701 (α-PsGal), and α-*N*-acetylgalactosaminidase from the marine bacterium *Arenibacter latericius* KMM 426^T (α-NaGa). All compounds are slow-binding irreversible inhibitors of α-PsGal, but have no effect on α-NaGa. A competitive inhibitor D-galactose protects α-PsGal against the inactivation. The inactivation rate (k_{inact}) and equilibrium inhibition (K_i) constants of monanchomycalin B, monanchocidin A, and normonanchocidin A were 0.166 ± 0.029 min^{-1} and 7.70 ± 0.62 μM, 0.08 ± 0.003 min^{-1} and 15.08 ± 1.60 μM, 0.026 ± 0.000 min^{-1}, and 4.15 ± 0.01 μM, respectively. The 2D-diagrams of α-PsGal complexes with the guanidine alkaloids were constructed with "vessel" and "anchor" parts of the compounds. Two alkaloid binding sites on the molecule of α-PsGal are shown. Carboxyl groups of the catalytic residues Asp451 and Asp516 of the α-PsGal active site interact with amino groups of "anchor" parts of the guanidine alkaloid molecules.

Keywords: sponge *Monanchora pulchra*; pentacyclic guanidine alkaloids; GH36 α-galactosidase; GH109 α-*N*-acetylgalactosaminidase; slow-binding irreversible inhibitor; monanchomycalin B; monanhocidin A; normonanhocidin A

---

1. Introduction

O-glycoside hydrolases are involved in the degradation of various poly- and oligosaccharides that serve as a source of carbon and energy for organism's growth, as well as performing various functions in organisms. Modification or blocking of these functions by powerful selective inhibitors underlies the treatment of a number of infectious diseases, malignant tumors and genetic disorders [1]. Inhibitors of enzymes are molecules that reduce or completely block the catalytic activity of an enzyme, causing either complete death of a cell or modification in the metabolic pathways. The marine sponges are important sources of enzyme inhibitors [2,3].

α-D-galactosidases (α-D-galactoside galactohydrolases, EC 3.2.1.22) catalyze the hydrolysis of non-reducing terminal α-D-galactose (Gal) from α-D-galactosides, galactooligosaccharides and polysaccharides. α-D-galactosidases are widespread among terrestrial plants, animals, human organs and tissues, as well as microorganisms [4]. The enzymes occur frequently in marine bacteria, especially in γ-Proteobacteria and Bacteroidetes [5–8]. The marine enzyme α-D-galactosidase (α-PsGal) was isolated from the cold-adaptable marine bacterium *Pseudoalteromonas* sp. KMM 701 inhabiting in the cold water of the Sea of Okhotsk [9]. The enzyme attracted our attention due to its ability to reduce the serological activity of B red blood cells [9]. The enzyme also interrupted the adhesion of *Corynebacterium diphtheria* to buccal epithelium cells at the neutral pH [10], and regulated the growth of biofilms of some pathogenic bacteria [11]. According to the carbohydrate active enzymes' classification (CAZy) [12] that is based on the amino acid sequence, α-PsGal belongs to the glycoside hydrolases (GHs) family 36.

α-*N*-Acetylgalactosaminidases (EC 3.2.1.49) catalyze the hydrolysis of the terminal α-linked *N*-acetylgalactosamine residues from the non-reducing ends of various complex carbohydrates and glycoconjugates. In the marine environment, α-*N*-acetylgalactosaminidases have been isolated from the liver and digestive organs of marine invertebrates, fishes, and marine bacteria of the genus *Arenibacter* [13,14]. The α-*N*-acetylgalactosaminidase from the marine bacterium *Arenibacter latericius* KMM 426^T (α-NaGa) was successfully applied for the complete conversion of A- into *O*-erythrocytes [15]. According to the CAZy classification, α-NaGa belongs to the GH109 family and is NAD^+-dependent *O*-glycoside hydrolase as α-*N*-acetylgalactosaminidase from a clinical pathogen *Elizabethkingia meningoseptica* [14]. Thus, α-PsGal and α-NaGa were found to be potential tools for blood transfusion as well as for structural studies in glycobiology and infection diseases. Therefore, screening and studying the natural inhibitors of these enzymes should be helpful for understanding the molecular machinery of their function and designing a method for their removal from the reaction for medical purposes.

Some of the secondary metabolites from marine sponges, which are biologically active compounds, were found to be applicable for pharmacology as the inhibitors of different classes for enzymes [3,16]. To date, a number of alkaloids of unique structures have been isolated from the marine sponge *Monanchora* sp. [17]. Their antitumor activity and mechanism of action have been shown [18–22]. The effect of pentacyclic guanidine alkaloids monanchomycalin B, monanchocidin A and normonanchocidin A isolated from the marine sponge *Monanchora pulchra* on the activity of exo-β-1,3-D-glucanases from the marine filamentous fungus *Chaetomium indicum* and endo-β-1,3-D-glucanase LIV from the marine bivalve mollusk *Spisula sachalinensis* was investigated [23]. In the present study, we focus our attention on the effect of the marine sponge secondary metabolites with a good therapeutic potential on the activity of two well-characterized α-glycosidases to elucidate the mechanism of their inhibitor action.

The present article aimed to compare of the effects of monanchomycalin B, monanchocidin A and normonanchocidin A on the activities of recombinant α-galactosidase from the marine bacterium *Pseudoalteromonas* sp. KMM 701 of the GH36 family and α-*N*-acetylgalactosaminidase from the marine bacterium *Arenibacter latericius* KMM 426^T of the GH109 family.

2. Results and Discussion

2.1. Identification of the Compounds

The samples of the marine sponge *M. pulchra* were collected in the Sea of Okhotsk (Kuril Islands region). The ethanol (EtOH) extract of the sponge *M. pulchra* sample N 047-243 was concentrated. The ethanol-soluble materials were further subjected to flash column chromatography on YMC*GEL ODS-A and high-performance liquid chromatography (HPLC) to obtain the pure monanchomycalin B (1). Monanhocidin A (**2**) and normonanhocidin A (**3**) were isolated from the EtOH extract of the sponge *M. pulchra* sample N 047-28 by the same method. The structure of the compounds **1**, **2** and **3** were assigned through the comparison of their spectral data with those reported in the references [20–22], respectively. Structural formulas of the pentacyclic guanidine alkaloids are shown on Figure 1.

The compounds isolated from the sponge *M. pulchra* have the same "vessel" part and differ in the structure of the "anchor" part of the molecule. The "anchor" part is presented by spermidine residue in monachomycalin B (1), by the tetra-substituted morpholinone derivative in monanchocidin A (2), and by the monosubstituted diaminopropane in normonanchocidin A (3).

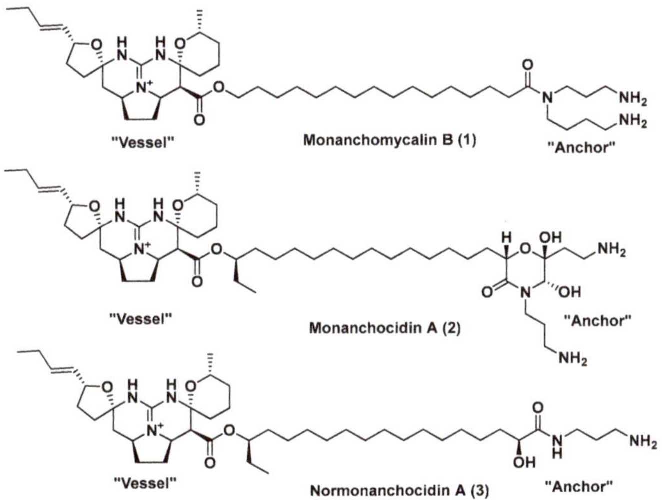

"Vessel" Monanchomycalin B (1) "Anchor"

"Vessel" Monanchocidin A (2) "Anchor"

"Vessel" Normonanchocidin A (3) "Anchor"

Figure 1. Structural formulas of pentacyclic guanidine alkaloids. "Vessel" part is on the left, and the "anchor" part is on the right of the molecule formula.

2.2. Effect of Monanchomycalin B, Monanchocidin A, and Normonanchocidin A on Activity of Two Glycosidases

The results of the pretreatment of two marine bacterial glycosidases with pentacyclic guanidine alkaloids within 30 min (Table 1) showed that all three compounds inhibited the activity of recombinant GH36 α-PsGal and had no effect on the recombinant GH109 α-NaGa.

Table 1. Residual activity v/v_0 (%) of the glycosidases after incubation with monanchomycalin B, monanchocidin A, or normonanchocidin A [1].

Enzyme	H_2O	Monanchomycalin B	Monanchocidin A	Normonanchocidin A
α-PsGal	100	0.21	2.7	1.7
α-NaGa	100	101.5	98.5	83.0

[1] Concentration of compounds in each sample was 0.2 mM, the enzyme was preincubated with an inhibitor for 30 min, 20 °C, pH 7.0.

It was previously shown that all three compounds significantly activated *Sps*LamIV endo-β(1→3)-D-glucanase of the mollusk *Spisula sachalinensis* and completely inhibited *Chin*Lam exo-β(1→3)-D-glucanase of the marine fungus *Chaetomium indicum* [23].

We have shown with the example of monanchomycalin B that pentacyclic guanidine alkaloids irreversibly inactivate the α-PsGal. The activity of the enzyme did not recover after dialysis against the buffer solution for 72 h (Table 2). The decrease of free enzyme activity by 2.6 times was observed, probably, due to the enzyme α-PsGal thermolability [24] and instability at the low concentrations (data not shown).

The study of the inhibitory effect of monanchomycalin B, monanchocidin A, and normonanhocidin A at different concentrations and incubation times showed that the IC_{50} values of compounds decreased with increasing of the incubation time of α-PsGal with inhibitors (data not shown). The results of kinetic studies on the α-PsGal inactivation by pentacyclic guanidine alkaloids are shown on Figure 2. The curves of the dependences of the residual activity v/v_0 on the time in semilogarithmic coordinates are shown in Figure 2a,c,e.

Table 2. The activity of α-PsGal after treating with monanchomycalin B [1].

Monanchomycalin B (μM)	Residual Activity (%)	
	Before Dialysis	After Dialysis
0	100	38
18	0	0

[1] The activity of α-PsGal after dialysis (72 h, 4 °C, 0.02 M sodium phosphate buffer (pH 7.0)) is presented considering the dilution of the enzyme sample. The results are average of three parallel measurements.

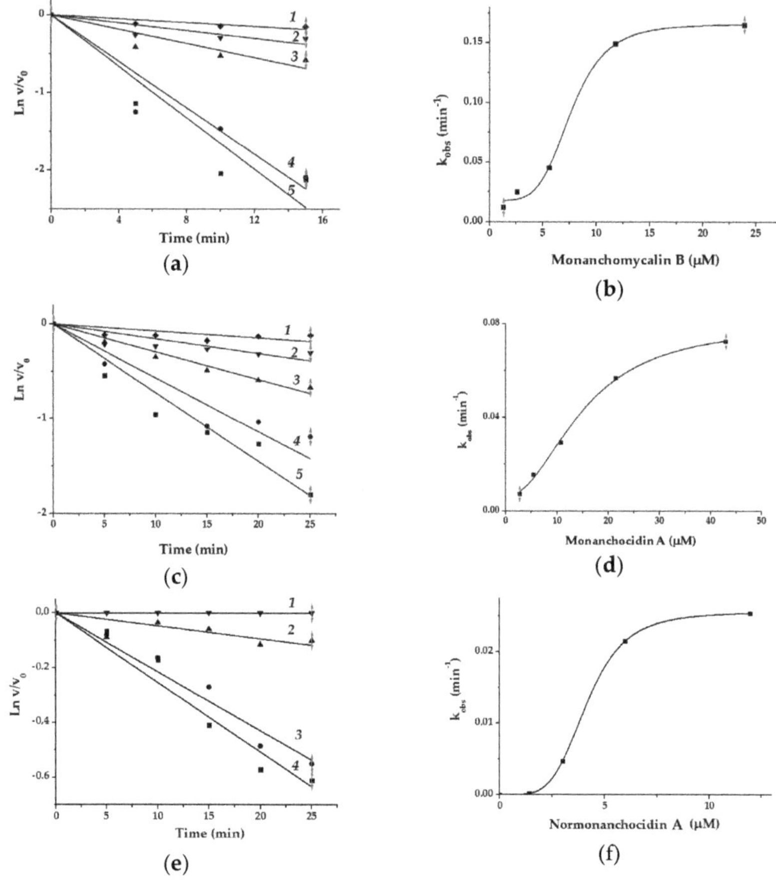

Figure 2. The results of kinetic studies of the α-PsGal inactivation by pentacyclic guanidine alkaloids: (**a**) the kinetic change of the residual activity of the enzyme (v/v_0) in semilogarithmic coordinates at 1.3 μM (1), 2.66 μM (2), (3) 5.69 μM, (4) 11.8 μM, and (5) 23.9 μM of monanchomycalin B; (**b**) the inactivation rate constants (k_{obs}) dependence on the concentrations of monanchomycalin B; (**c**) the kinetic change of the residual activity of the enzyme in semilogarithmic coordinates at 2.7 μM (1), 5.4 μM (2), (3) 10.7 μM, (4) 21.4 μM, and (5) 42.9 μM of monanchocidin A; (**d**) the inactivation rates (k_{obs}) dependence on the concentrations of monanchocidin A; (**e**) the kinetic change of the residual activity of the enzyme in semilogarithmic coordinates at 1.49 μM (1), 2.98 μM (2), 5.97 μM (3), and 11.9 μM (4) of normonanchocidin A; (**f**) the inactivation rates (k_{obs}) dependence on the concentrations of normonanchocidin A. All of the experiments were performed in duplicates.

The α-PsGal inactivation developed relatively slowly, within a few minutes under these experimental conditions. In this case, the inhibitory activity of the compounds can be more accurately described by the inactivation rate constant (k_{inact}, min^{-1}) and equilibrium inhibition constant K_i [25]. The values of k_{obs} increased together with the compound concentrations. Sigmoid curves of k_{obs} dependences on concentration of the inhibitors (Figure 2b,d,f) mean that the process of the enzyme (E) inactivation by slowly-binding irreversible inhibitors (I) has a cooperative character, and occurs in two stages: (i) the formation of a reversible enzyme-inhibitor complex [E I^n] and (ii) irreversible inactivation of the enzyme in the E-I^n complex. The kinetic Equation (1) describes the irreversible slow inhibition of α-PsGal under the action of the pentacyclic guanidine alkaloids:

$$E + nI \overset{K_i}{\rightleftharpoons} [E\ I^n] \overset{k_{inact}}{\rightarrow} E\text{-}I^n, \quad (1)$$

where n is coefficient of cooperativity, which is interpreted as the number of identical binding sites; K_i is an equilibrium constant of inhibition (μM). The experimental dependences of k_{obs} on the concentration of the compounds (I) (Figure 2b,d,g) are approximated by the Hill's Equation (2).

$$k_{obs} = k_{inact}\ I^n/(K_i^{\ n} + I^n), \quad (2)$$

The results of the experimental data fitting with theoretical curves are shown in Table 3.

Table 3. The α-PsGal inhibition constants for monanchomycalin B, monanchocidin A, and normonanchocidin A.

Inhibitor	k_{inact} (min^{-1})	K_i (μM)	n
Monanchomycalin B	0.166 ± 0.029	7.70 ±0.62	4.64 ± 1.21
Monanchocidin A	0.08 ± 0.003	15.08 ± 1.60	2.1 ± 0.47
Normonanchocidin A	0.026 ± 0.000	4.15 ± 0.01	4.55 ± 0.02

According to the values of K_i and k_{inact}, the alkaloids can be arranged in descending of binding-affinity as normonanchocidin A > monanchomycalin B > monanchocidin A, but in increasing inactivation rate in the following order: monanchomycalin B > monanchocidin A> normonanchocidin A.

Thus, based on the results of the kinetic studies, we have suggested that the pentacyclic guanidine alkaloids are slow-binding inhibitors for α-PsGal similarly to the *Chin*Lam glucanase [23]. It is accepted that slow-binding inhibition is observed whenever an enzyme-inhibitor complex forms or undergoes further conversion, at a slower rate relative to the overall reaction rate [26]. Inhibitors of peptidases [27], monoamine oxidases, and acetylcholinesterases [28,29] are examples of the slow-binding. Previously, the property of monanchocidin A as a slow-acting biologically active compound was shown for cancer cells [30]. For α-PsGal, the chlorine and bromine echinochrome derivatives from a sea urchin have been previously shown to be slow-binding inactivators as well [31]. Moreover, we have found that the inhibition rate increases with the binding of at least four molecules of the compounds.

D-galactose being a competitive inhibitor for α-galactosidases of GH36 family [32,33] decreased the activity of α-PsGal on 50% at 0.7 mM. The active-site-directed nature of the inactivation was proven by demonstration of the enzyme's protection against inactivation by D-galactose (Figure 3).

From the Figure 3, it is evident that this monosaccharide significantly protects α-PsGal from the inactivation.

Regardless of the inhibitor concentration, $k_{obs}{}^{Gal}$ decreased on average by 50% in the presence of the reaction product D-galactose (Table 4). The monosaccharide partially protects the enzyme from inactivation. This suggests that the inhibitor interacts with the enzyme molecule in the region of the active center.

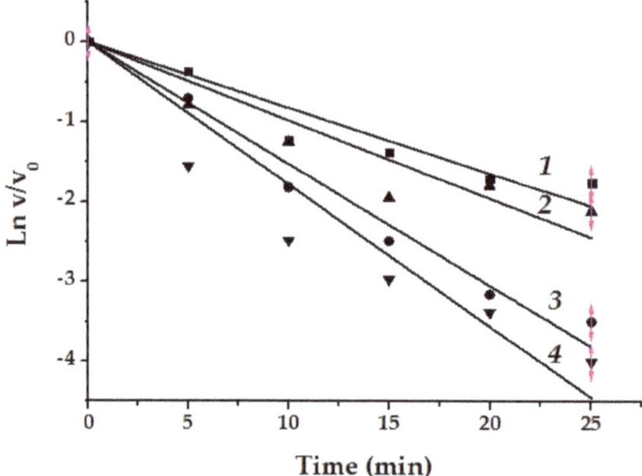

Figure 3. Protection of α-PsGal activity by D-galactose (0.7 mM) against monanchomycalin B inactivation: curves 1 and 2 show the effect of the enzyme activation rate on incubation time with the inhibitor (11.4 μM and 14.2 μM, respectively) in the presence of D-galactose in semi-log coordinates; curves 3 and 4 represent the rates of enzyme inactivation at the same inhibitor concentration and incubation time without D-galactose. All of the experiments were performed in duplicates.

Table 4. Protection of α-PsGal activity by D-galactose against monanchomycalin B inactivation.

Concentration (μM)	k_{obs} (min^{-1})	k_{obs}^{Gal} (min^{-1}) [1]
11.4	0.152 ± 0.005	0.082 ± 0.006
14.2	0.178 ± 0.013	0.097 ± 0.008

[1] k_{obs}^{Gal}—in the presence of D-galactose.

Taking into account that the active center of the enzyme and the "vessel" part of the molecules of the compounds are identical in all the experiments, their inhibitory properties towards α-PsGal are determined by the structure of the "anchor" part. In this case, the diaminopropane residue has the greatest affinity, but more slowly penetrates to the active center of the GH36 α-PsGal from the marine bacterium. The monosubstituted diaminopropane has been shown to be also the best inhibitor for the *Chin*Lam exo-(1→3)β-D-glucanase from a marine fungus as well [23]. However, these compounds did not show inhibitor properties towards the GH109 α-NaGa from the marine bacterium *A. latericius* as well as the GH16 endo-(1→3)β-D-glucanase from the marine bivalve mollusk *S. sachalinensis* [23].

2.3. Theoretical Model of the Guanidine Alkaloids Complexes with α-Galactosidase

The enzyme α-PsGal is a typical *O*-glycoside hydrolase of the GH36 family. It was previously shown that its molecule consists of two identical subunits [9,10]. One subunit is a three-domain protein. The active center is located in the central $(\beta/\alpha)_8$ domain. Asp 451 and Asp 516 are catalytic residues [24].

Figure 4 shows 2D-diagrams of the α-PsGal complexes with the guanidine compounds. The "vessel" part identical for the all compounds (Figure 4a), and the "anchor" parts of monanchomycalin B (Figure 4b), monanchocidin A (Figure 4c), and normonanchocidin A (Figure 4d) complexes with the active center of α-PsGal were built by molecular docking of the program Molecular Operating Environment version 2018.01 (MOE) [34].

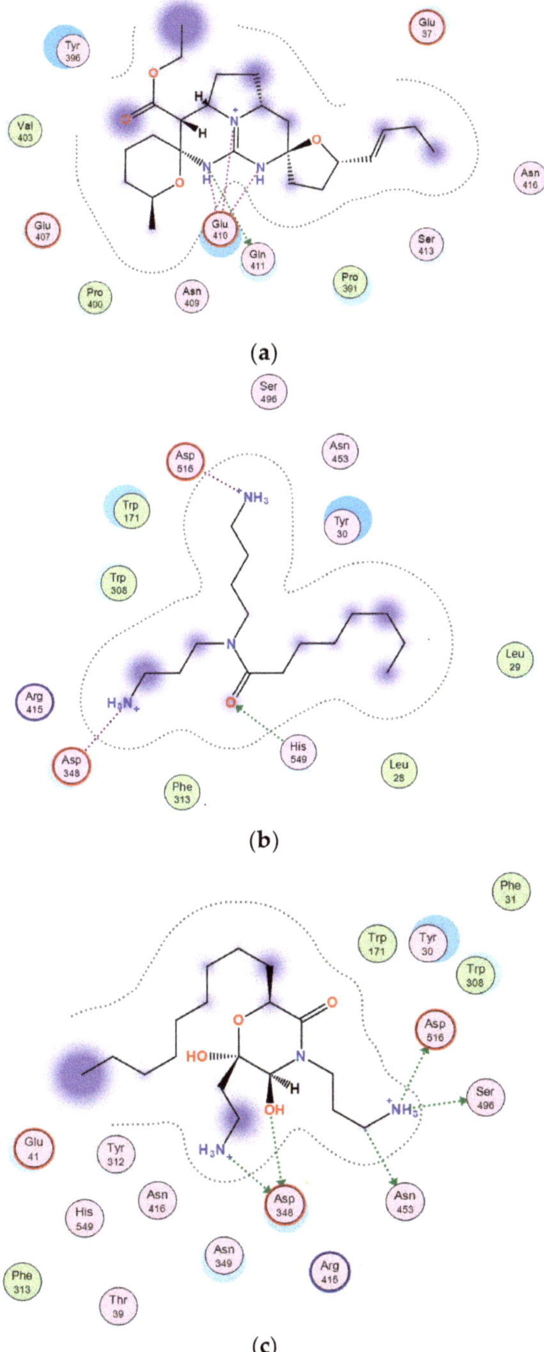

Figure 4. *Cont.*

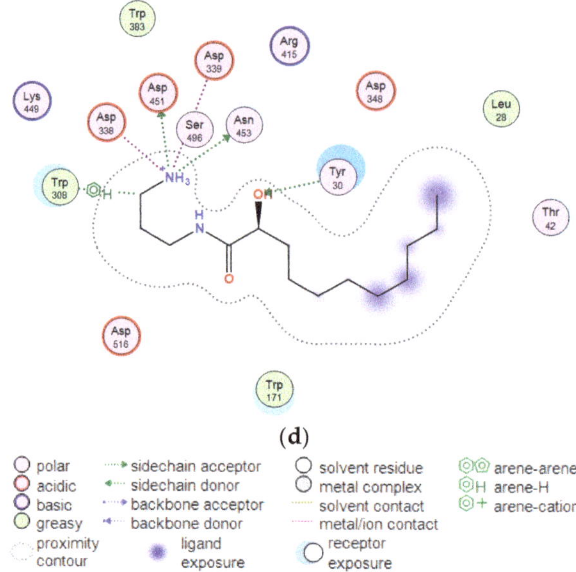

Figure 4. 2D-diagrams of the α-PsGal complexes with the guanidine alkaloids: (**a**) 2D-diagram of α-PsGal—"vessel" part complex; (**b**) 2D-diagram of the α-PsGal-spermidine residue of monachomycalin B; (**c**) 2D-diagram of the α-PsGal-tetra-substituted morpholinone derivative of monanchocidine A; (**d**) 2D-diagram of α-PsGal—the monosubstituted diaminopropane of normonanchocidine A.

The "anchor" parts of the compounds are the spermidine residue in monachomycalin B, tetra-substituted morpholinone derivative in monanchocidine A, and monosubstituted diaminopropane in normonanchocidine A.

Two different binding sites for the "vessel" and "anchor" parts of alkaloids in the molecule of α-PsGal were found. The carboxyl groups of the catalytic residues Asp451 and Asp516 in the active site of α-PsGal take part directly in the interaction with amino groups of "anchor" parts of the compounds.

The molecules of the test compounds consist of two polar nitrogen-containing residues connected by hydrophobic polymethylene chains. In this case, the "anchor" part of the molecule is very mobile. Based on the simulation results, the "vessel" part of the molecule binds near the crater of the active center and does not influence the activity of the enzyme, but directs and promotes an increase in the affinity of the "anchor" part; thus, the binding of the latter occurs more slowly and leads to the loss of enzyme activity. D-galactose located in the active center prevents the spermidine residue of monachomycalin B from entering to the catalytic site what can slow down the inactivation of the enzyme (Figure S1). In accordance with the results of a 3D-superposition of the α-PsGal active site with D-galactose and anchor parts of the compounds, the monosubstituted diaminopropane of normonanchocidine A penetrates most deeply into the pocket of the active center of α-PsGal (Figure S1c).

3. Materials and Methods

3.1. Materials

The 4-nitrophenyl-α-D-galactopyranoside (pNP-α-Gal), D-galactose (Gal), and 4-nitrophenyl-α-*N*-acetylgalactosaminide (pNP-α-NAcGal) were purchased from Sigma-Aldrich Chemical Company (St. Louis, MO, USA). Encyclo DNA-polymerase and enterokinase were purchased from Evrogen JSC (16/10 Miklukho-Maklaya str., Moscow, Russian Federation). Nco I and Sal I were purchased from New England Biolabs (NEB), Ipswich, MA, USA. pET 40 b(+) plasmid was purchased from Invitrogen, Carlsbad, CA, USA. Bacto-tryptone, sorbitol, $MgCl_2$, KH_2PO_4, kanamycin, glycerin phenyl methanesulfonyl fluoride PMSF were purchased from (Helicon, Moscow, Russian Federation, Kutuzovsky Prospect, 88). Sodium phosphates one- and two-substituted were purchased from PanReac AppliChem GmbH (Ottoweg 4, Darmstadt, Germany). IMAC Ni^{2+} Sepharose, Q-Sepharose, Mono-Q, and Superdex-200 PG were purchased from GE Healthcare (Uppsala, Sweden). EtOH, trifluoroacetic acid (TFA), CD_3OD were from the Russian Federation.

3.2. Experimental Equipment

Optical rotation was measured on Perkin-Elmer 343 polarimeter (Waltham, MA, USA). The ^1H and ^{13}C nuclear magnetic resonance (NMR) spectra were recorded on Avance III-700 spectrometer (Bruker BioSpin GmbH, Silberstreifen 4, D-76287 Rheinstetten/KarlsruheIpswich, Germany) at 700 and 175 MHz, respectively. The chemical shifts were correlated in accordance with the CD_3OD signals (δ_H 3.30/δ_C 4 9.60). Electrospray ionization (ESI) mass spectra (including HRESIMS) were measured using a Bruker Impact II Q-TOF mass spectrometer (Bruker Daltonics, Bremen, Germany). HPLC was performed using Shimadzu instrument (Shimadzu Corporation, Kyoto, Japan) with a diffraction refractometer RID-DE14901810 and YMC-ODS-A column (YMC CO., LTD., Kyoto, Japan). Microplate spectrophotometer (BioTek Instruments, Highland Park, Winooski, VT, USA) was used for measuring of optical density at 400 nm (D_{400}).

3.3. Collection and Identification of Sponge Material

Samples of the sponge *Monanchora pulchra* N 047-028 and N 047-243 were collected by the dredging method during the 47th scientific cruise of the R/V *Akademik Oparin* in August 2015 at the Chirpoi (49°24.1 N, 154°17.8 E, depth of 139 m) and Onekotan, the Kuril Islands (49°24.1 N and 154°17.8 E, depth of 135 m). Identification of sponges was performed by V.B. Krasokhin. The voucher specimens are kept in the collection of G.B. Elyakov Pacific Institute of Bioorganic Chemistry, Far Eastern Branch, Russian Academy of Sciences (www.piboc.dvo.ru).

3.4. Isolation and Purification of Compounds

The freshly collected *M. pulchra* samples (N 047-243 and N 047-28) were extracted with EtOH and a part of which (30 mL) was concentrated in vacuo. The residual part was chromatographed on a microcolumn (10 × 12 mm) with YMC*GEL ODS-A reversed-phase sorbent (75 μm) using aqueous EtOH (40%), and then EtOH (65%)–H_2O (35%)–TFA (0.1%). The eluates with TFA were evaporated. The compounds were isolated by HPLC using YMC-ODS-A column (250 × 10 mm) and EtOH (65%)–H_2O (35%)–TFA (0.1%) to afford pure compound **1** (4.0 mg) from the sample 047-243, as well as compounds **2** (0.8 mg), and **3** (0.8 mg) from the sample 047-28.

Monanchomycalin B (**1**): high-resolution electrospray ionisation mass spectrometry (HRESI MS) *m/z* 785.6259 $[M + H]^+$, (calcd. for $C_{45}H_{81}N_6O_5$: 785.6263);

Monanchocidin A (**2**): HRESI MS *m/z* 859.6267, $[M + H]^+$, (calcd. for $C_{47}H_{83}N_6O_8$: 859.6267);

Normonanchocidin A (**3**): HRESI MS *m/z* 758.5792, $[M + H]^+$, (calcd. for $C_{43}H_{76}N_5O_6$: 758.5790).

3.5. Production of Recombinant Enzymes

3.5.1. Production and Purification of Recombinant α-D-galactosidase

The recombinant wild-type α-D-galactosidase α-PsGal was produced as described earlier [35]. The plasmid DNA pET-40b(+) containing insertion of the gene from the marine bacterium *Pseudoslteromonas* sp. KMM 701 encoding α-PsGal was transformed in the *Escherichia coli* strain Rosetta (DE3). Heterological expression was carried out at optimal conditions as described previously [36]. Purification of the recombinant α-PaGal was performed according to the procedures described in the reference [35].

3.5.2. Production and Purification of Recombinant α-Nacetylgalactosaminidase

The recombinant wild-type α-Nacetylgalactosaminidase α-NaGa was produced as described earlier [35]. The plasmid DNA pET-40b(+) containing insertion of the gene from the marine bacterium *Arenibacter latericius* KMM 426^T encoding α-NaGa was transformed in the *Escherichia coli* strain Rosetta (DE3). Heterological expression was carried out at optimal conditions as described previously [37].

The new purification procedure was modified and carried out at 4 °C. The cleared supernatant containing α-NaGa and 20% glycerol was loaded directly onto a Ni-sepharose column (5 cm × 36 cm) equilibrated with the buffer A (10 mM NaH_2PO_4, 10 mM Na_2HPO_4, 0.5 M NaCl, 5 mM imidazole, 20% glycerol, pH 8.0). The recombinant protein was eluted with the 5–500-mM linear imidazole gradient. The eluted fractions were analyzed, collected and dialyzed against the buffer B (10 mM NaH_2PO_4, 10 mM Na_2HPO_4, 50% glycerol, pH 8.0). Then, the protein solution was loaded onto a column (2 cm × 15 cm) with an ion-exchange resin Source 15Q equilibrated with the buffer B. The recombinant protein was eluted with the 0–1.5 M linear NaCl gradient. The fractions exhibiting the activity of α-NaGa were collected and examined using sodium dodecyl sulphate polyacrylamide gel electrophoresis (SDS-PAGE).

3.5.3. Enzyme and Protein Assays

The activity of α-PsGal and α-NaGa were determined by increasing the amount of p-nithrophenol (pNP). The mixtures containing 50 μL of an enzyme solution and 100 μL of a substrate solution (1 mg/mL) in 0.05 M sodium phosphate buffer (pH 7.0) were incubated at 20 °C during 5 min for α-PsGal and 30 min for α-NaGa. The reactions were stopped by the addition of 150 μL of 1 M Na_2CO_3. One unit of the activity (U) was determined as the amount of an enzyme that releases 1 μmol of pNP per 1 min at 20 °C. The amount of the released pNP was determined spectrophotometrically (ε_{400} = 18,300 M^{-1} cm^{-1}). The specific activity was calculated as U/mg of protein. The protein concentration was determined by the Bradford method calibrated with BSA as a standard [38]. Buffer solutions of α-PsGal (0.1 U/mL) and α-NaGa (0.05 U/mL) were used in the further experiments.

3.6. *Effects of Pentacyclic Guanidine Alkaloids on Glycosidases of Marine Bacteria*

3.6.1. The Effects of Monanchomycalin B, Monanchocidin A, and Normonanchocidin A on Glyctosidases

To study the effect of pentacyclic guanidine alkaloids on α-PsGal and α-NaGa, 25 μL of an aqueous solution of monanchomycalin B, monanchocidin A, or normonanchocidin A (1 mg/mL) was mixed with 50 μL of the enzyme solutions in wells of the 96-cell plates and incubated for 30 min. Reactions were initiated by the addition of 75 μL of substrate solutions (p-nithrophenil galactopyranoside (pNP-α-Gal) for α-PsGal and p-nithrophenil N-acetylgalactosaminide (pNP-α-NAcGal) for α-NaGa) in 0.05 M sodium phosphate buffer (pH 7.0). The reaction mixture was incubated at 20 °C for 2–30 min in the final volume 150 μL, and then 150 μL of 1M Na_2CO_3 solution was added to the incubation mixture to stop the reaction. Each reaction mixture was prepared in duplicate. The absorbance was measured at 400 nm. Results were read with a Gen5 and treated with ExCel software. The activity of α-PsGal or α-NaGa was determined as described above. The residual activity was calculated as the ratio v/v_0 (%), where v is the enzyme activity in the presence of an inhibitor, and v_0 is the enzyme activity in the absence of an inhibitor. The v_0 was taken for 100%.

3.6.2. The Irreversibility of Monanchomycalin B Inhibition

To determine the reversibility of the inhibition of the α-PsGal activity, 40 μL (18 μM, H_2O) of the monanchomycalin B solution was added to 60 μL of the enzyme solution; the mixture was incubated for 60 min. Two volumes of 20 μL were taken from the reaction mixture, 380 μL of a pNP-α-Gal solution (3.32 mM, in probe was ~5 K_m) was added to each mixture, and then the reaction was stopped by addition of 0.6 mL of 1M Na_2CO_3 after 30 min of incubation. The value of optical density at the wavelength 400 (OD_{400}) was measured in the 1-cm cuvette. The activity of α-PsGal was determined as described above. The remaining 60 μL of the reaction mixture was dialyzed against 1 L of 0.02 M sodium phosphate buffer (pH 7.0) for 72 h at 4 °C. To estimate the dilution, the volume of the reaction mixture after dialysis was measured. The enzyme activity was determined as described above and recalculated taking the dilution into account (1.7 times). A sample of α-PsGal untreated by monanchomycalin B (60 μL of the enzyme solution and 40 μL of H_2O) was used as a control. The experiment was carried out in two replicates. The residual activity was calculated as described above.

3.6.3. The Assay of α-PsGal Inhibition by Monanchomycalin B, Monanchocidin A, and Normonanchocidin A

For the α-PsGal inhibition assay, 50 μL of the enzyme solution in 0.05 M sodium phosphate buffer (pH 7.0) were placed in the cells of the 96-well plate with 10 μL of a compound water solution at various concentrations in probes (186, 92.8, 46.4, 23.2, 11.6, 5.8, 2.9, 1.3, 0 μM were for monanchomycalin B; 171.5, 85.7, 42.9, 21.4, 10.7, 5.4, 2.7, 0 μM were for monanchocidin A, and 191.1, 95.6, 47.8, 23.9, 11.9, 6.0, 3.0, 1.5, 0 μM were for normonanchocidin A), and incubated for each concentration during 5, 10, 15, 20, and 25 min. The enzyme reaction was initiated by the addition of 90 μL of the pNP-α-Gal solution (3.32 mM, ~5 K_m for probe) in 0.05 M sodium phosphate buffer (pH 7.0). The reaction mixtures were incubated for 2 - 15 min in the final volume 150 μL, then 150 μL of the water solution of Na_2CO_3 (1 M) were added to stop the reaction, and OD_{400} for the reaction mixtures were immediately measured by a microplate spectrophotometer. The time of each reaction was strictly monitored by stopwatch. The standard and residual activity v/v_0 were calculated as described above.

3.6.4. The Kinetic Parameters of Inactivation

The equilibrium inhibition constants (K_i) and kinetic inactivation constants (k_{inact}) were determined by the classical methods [39]. The inactivation of α-PsGal by the different concentrations of inhibitors (1.3–50 μM) was performed in 0.05 M sodium phosphate buffer (pH 7.0) at a temperature 20 °C. An aqueous solution (25 μL) of the compound at the different concentrations was added to 50 μL of the α-PsGal solution (0.2 U/mL), held for 5, 10, 15, 20, and 25 min at 20 °C, then 100 μL of the pNP-α-Gal solution was added and incubated for 2–15 min at 20 °C. The same conditions were used in the control reaction, but the inhibitor was replaced with distilled water. The reactions were stopped by the addition of 1 M Na_2CO_3 (150 μL); the amount of pNP formed in 1 min was determined as described above. The residual activity v/v_0 was presented as a function of time. The pseudo-first-order rate constant of inactivation (k_{obs}) was determined for each inactivator concentration as the slope of the v/v_0 dependence on the incubation time in semilogarithmic coordinates. The ExCel software was used for these calculations. The second order rate constants for the inactivation process were determined by fitting the dependences of the k_{obs} values on the concentration of the inactivators to the Hill's equations. An analysis of the curves and the choice of models for calculation of K_i (μM) and k_{inact} (min^{-1}) were performed with the Origin 8.1 software (OriginLab, Northampton, MA, USA).

3.6.5. Protection of α-PsGal Inactivation by D-galactose

The active-site-directed nature of the inactivation was confirmed by demonstrating protection against the inactivation by competitive inhibitor D-galactose. Inactivation mixtures (75 μL) containing 50 μL of the enzyme solution and 10 μL of D-galactose (0.7 mM in mixture) were preincubated for 15 min, then 15 μL of the monanchomycalin B solution (11.4 μM and 14.2 μM in mixture) were added and incubated at various time intervals as described above. The residual activity of the enzyme was assayed as described above.

3.7. Theoretical Models of α-PsGal Complexes with Guanidine Alkaloids

The target-template alignment customization of the modeling process and 3D model building of α-PsGalA (GenBank: ABF72189.2) were carried out using the Molecular Operating Environment version 2018.01 [37] package using the forcefield Amber12: EHT. The α-D-galactosidase from *Lactobacillus acidophilus* NCFM (Protein data bank (PDB) code: 2XN2) with a high-resolution crystal structure was used as a template. The evaluation of structural parameters, contact structure analysis, physical-chemical properties, molecular docking, and visualization of the results were carried out with the Ligand interaction and Dock modules in the MOE 2018.01 program. The results were obtained using the equipment of Shared Resource Center Far Eastern Computing Resource of Institute of Automation and Control Processes Far Eastern Branch of the Russian Academy of Sciences (IACP FEB RAS) [40].

4. Conclusions

For study the effect of the marine sponge metabolites with a therapeutic potential, we used two well-characterized α-glycosidases for justifying a possible mechanism of their inhibitor action. Monanchomycalin B, normonanchocidin A, monanchocidin A have been shown to be irreversible slow-binding inhibitors of the GH36 family α-galactosidase α-PsGal from the marine bacterium *Pseudoalteromonas* sp. KMM 701, but have no effect on the activity of the GH109 family α-NaGa from the marine bacterium *Arenibacter latericius* KMM 426^{T}. The inhibitory ability of the alkaloids depends on the chemical structure of the anchor parts of their molecules. The alkaloids can be arranged in the descending order of the binding-affinity: normonanchocidin A > monanchomycalin B > monanchocidin A, and in the decreasing order of the inactivation rate: monanchomycalin B > monanchocidin A > normonanchocidin A. These highly active marine compounds selectively acted on the enzymes from the different structural GH families, binding to the electronegative areas of the protein surfaces formed mainly by carboxylic acid side groups in the active-site-directed manner. The well-characterized α-glycosidases of marine bacteria have been proved to be suitable models for characterizing the novel properties of the alkaloids.

Supplementary Materials: The following are available online at http://www.mdpi.com/1660-3397/17/1/22/s1, Figure S1. 3D-superposition of the α-PsGal active site with D-galactose and spermidine residue of monachomycalin B (**a**), tetra-substituted morpholinone derivative of monanchocidine A (**b**) and monosubstituted diaminopropane of normonanchocidine A (**c**). Parts of guanidine alkaloids are shown as "ball and stick" with grey color, galactose shown as "stick" with yellow color. The molecular surface closed to the ligands is shown in pink (H-bonding), green (hydrophobic) and blue (mild polar).

Author Contributions: Conceptualization, enzyme investigation, and writing-original draft preparation: I.B.; bioinformatics analysis and computer modeling of protein-inhibitor complex structures: G.L.; construction, expression, and purification of recombinant and mutant enzymes: L.S. and L.B.; purification and identification of compounds: L.Sh. and T.M.; formal analysis: O.S.; resources: L.T.

Funding: This research was funded by the framework of the State Assignment of the Pacific Institute of Bioorganic Chemistry, Far Eastern Branch, Russian Academy of Sciences, project no. 0266-2016-0002. The financial support is provided by Ministry of Education and Science of Russia (Agreement 02.G25.31.0172, 01.12.2015). Isolation and identification of alkaloids were performed with a partial support of the Far Eastern Branch of the Russian Academy of Sciences, project no. 18-4-026.

Conflicts of Interest: The authors declare no conflict of interest.

References

1. Asano, N. Glycosidase inhibitors: Update and perspectives on practical use. *Glycobiology* **2003**, *13*, 93R–104R. [CrossRef] [PubMed]
2. Skropeta, D.; Pastro, N.; Zivanovic, A. Kinase inhibitors from marine sponges. *Mar. Drugs* **2011**, *9*, 2131–2154. [CrossRef] [PubMed]

3. Ruocco, N.; Costantini, S.; Palumbo, F.; Constantini, M. Marine sponges and bacteria as challenging sources of enzyme inhibitors for pharmacological applications. *Mar. Drugs* **2017**, *15*, 173. [CrossRef] [PubMed]
4. Bakunina, I.Y.; Balabanova, L.A.; Pennacchio, A.; Trincone, A. Hooked on α-D-galactosidases: From biomedicine to enzymatic synthesis. *Crit. Rev. Biotechnol.* **2016**, *36*, 233–245. [CrossRef] [PubMed]
5. Bakunina, I.Y.; Ivanova, E.P.; Nedashkovskaya, O.I.; Gorshkova, N.M.; Elyakova, L.A.; Mikhailov, V.V. Search for α-D-galactosidase producers among marine bacteria of the genus *Alteromonas*. *Prikl. Biokh. Mikrobiol. (Moscow)* **1996**, *32*, 624–628.
6. Ivanova, E.P.; Bakunina, I.Y.; Nedashkovskaya, O.I.; Gorshkova, N.; Mikhailov, V.V.; Elyakova, L.A. Incidence of marine microorganisms producing β-*N*-acetylglucosaminidases, α-D-galactosidases and α-*N*-acetylgalactosaminidases. *Rus. J. Mar. Biol.* **1998**, *24*, 365–372.
7. Bakunina, I.Y.; Nedashkovskaya, O.I.; Alekseeva, S.A.; Ivanova, E.P.; Romanenko, L.A.; Gorshkova, N.M.; Isakov, V.V.; Zvyagintseva, T.N.; Mikhailov, V.V. Degradation of fucoidan by the marine proteobacterium *Pseudoalteromonas citrea*. *Microbiology (Moscow)* **2002**, *71*, 41–47. [CrossRef]
8. Bakunina, I.Y.; Nedashkovskaya, O.I.; Kim, S.B.; Zvyagintseva, T.N.; Mihailov, V.V. Diversity of glycosidase activities in the bacteria of the phylum *Bacteroidetes* isolated from marine algae. *Microbiology (Moscow)* **2012**, *81*, 688–695. [CrossRef]
9. Bakunina, I.Y.; Sova, V.V.; Nedashkovskaya, O.I.; Kuhlmann, R.A.; Likhosherstov, L.M.; Martynova, M.I.; Mihailov, V.V.; Elyakova, L.A. α-D-galactosidase of the marine bacterium *Pseudoalteromonas* sp. KMM 701. *Biochemisrty (Moscow)* **1998**, *63*, 1209–1215.
10. Balabanova, L.A.; Bakunina, I.Y.; Nedashkovskaya, O.I.; Makarenkova, I.D.; Zaporozhets, T.S.; Besednova, N.N.; Zvyagintseva, T.N.; Rasskazov, V.A. Molecular characterization and therapeutic potential of a marine bacterium *Pseudoalteromonas* sp. KMM 701 α-D-galactosidase. *Mar. Biotechnol.* **2010**, *12*, 111–120. [CrossRef]
11. Terentieva, N.A.; Timchenko, N.F.; Balabanova, L.A.; Golotin, V.A.; Belik, A.A.; Bakunina, I.Y.; Didenko, L.V.; Rasskazov, V.A. The influence of enzymes on the formation of bacterial biofilms. *Health Med. Ecol. Sci.* **2015**, *60*, 86–93.
12. Lombard, V.; Golaconda Ramulu, H.; Drula, E.; Coutinho, P.M.; Henrissat, B. The Carbohydrate-active enzymes database (CAZy) in 2013. *Nucleic Acids Res.* **2014**, *42*, D490–D495. [CrossRef] [PubMed]
13. Bakunina, I.Y.; Nedashkovskaya, O.I.; Kim, S.B.; Zvyagintseva, T.N.; Mikhailov, V.V. Distribution of α-*N*-acetylgalactosaminidases among marine bacteria of the phylum *Bacteroidetes*, epiphytes of marine algae of the Seas of Okhotsk and Japan. *Microbiology (Moscow)* **2012**, *81*, 375–380. [CrossRef]
14. Bakunina, I.Y.; Nedashkovskaya, O.I.; Balabanova, L.A.; Zvyagintseva, T.N.; Rasskasov, V.A.; Mikhailov, V.V. Comparative analysis of glycoside hydrolases activities from phylogenetically diverse marine bacteria of the genus *Arenibacter*. *Mar. Drugs* **2013**, *11*, 1977–1998. [CrossRef] [PubMed]
15. Bakunina, I.Y.; Kuhlmann, R.A.; Likhosherstov, L.M.; Martynova, M.D.; Nedashkovskaya, O.I.; Mikhailov, V.V.; Elyakova, L.A. α-*N*-Acetylgalactosaminidase from marine bacterium *Arenibacter latericius* KMM 426^T removing blood type specificity of A erythrocytes. *Biochemistry (Moscow)* **2002**, *67*, 689–695. [CrossRef]
16. Blunt, J.W.; Carroll, A.R.; Copp, B.R.; Davis, R.A.; Keyzers, R.A.; Prinsep, M.R. Marine natural products. *Nat. Prod. Rep.* **2018**, *35*, 8–53. [CrossRef] [PubMed]
17. Liu, J.; Li, X.-W.; Guo, Y.-W. Recent advances in the isolation, synthesis and biological activity of marine guanidine alkaloids. *Mar. Drugs* **2017**, *15*, 324. [CrossRef]
18. Sfecci, E.; Lacour, T.; Amade, P.; Mehiri, M. Polycyclic guanidine alkaloids from Poecilosclerida Marine Sponges. *Mar. Drugs* **2016**, *14*, 77. [CrossRef]
19. Shrestha, S.; Sorolla, A.; Fromont, J.; Blancafort, P.; Flematti, G.R. Crambescidin 800, isolated from the marine sponge *Monanchora viridis*, induces cell cycle arrest and apoptosis in triple-negative breast cancer cells. *Mar. Drugs* **2018**, *16*, 53. [CrossRef]
20. Guzii, A.G.; Makarieva, T.N.; Denisenko, V.A.; Dmitrenok, P.S.; Kuzmich, A.S.; Dyshlovoy, S.A.; Krasokhin, V.B.; Stonik, V.A. Monanchocidin: A new apoptosis-inducing polycyclic guanidine alkaloid from the marine sponge *Monanchora pulchra*. *Org. Lett.* **2010**, *12*, 4292–4295. [CrossRef]

21. Makarieva, T.N.; Tabakmaher, K.M.; Guzii, A.G.; Denisenko, V.; Dmitrenok, P.S.; Kuzmich, A.S.; Lee, H.-S.; Stonik, V.A. Monanchomycalins A and B, unusual guanidine alkaloids from the sponge *Monanchora pulchra*. *Tetrahedron Lett.* **2012**, *53*, 4228–4231. [CrossRef]
22. Tabakmakher, K.M.; Makarieva, T.N.; Denisenko, V.A.; Guzii, A.G.; Dmitrenok, P.S.; Kuzmich, A.S.; Stonik, V.A. Normonanchocidins A, B and D, new pentacyclic guanidine glkaloids from the Far-Eastern marine sponge *Monanchora pulchra*. *Nat. Prod. Commun.* **2015**, *10*, 913–916. [PubMed]
23. Dubrovskaya, Y.V.; Makarieva, T.N.; Shubina, L.K.; Bakunina, I.Y. Effect of pentacyclic guanidine alkaloids from the marine sponge *Monanchora pulchra* Lambe, 1894 on activity of natural β-1,3-D-glucanase from the marine fungus *Chaetomium indicum* Corda, 1840 and the marine bivalve mollusk *Spisula sachalinensis*, Schrenck, 1861. *Rus. J. Mar. Biol.* **2018**, *44*, 127–134.
24. Bakunina, I.; Slepchenko, L.; Anastyuk, S.; Isakov, V.; Likhatskaya, G.; Kim, N.; Tekutyeva, L.; Son, O.; Balabanova, L. Characterization of properties and transglycosylation abilities of recombinant α-galactosidase from cold adapted marine bacterium *Pseudoalteromonas* KMM 701 and its C494N, D451A mutants. *Mar. Drugs* **2018**, *16*, 349. [CrossRef] [PubMed]
25. Parsons, Z.D.; Gates, K.S. Redox regulation of protein tyrosine phosphatases: Methods for kinetic analysis of covalent enzyme inactivation. *Methods Enzymol.* **2013**, *528*, 129–154. [PubMed]
26. Morrison, J.F. The slow-binding and slow, tight-binding inhibition of enzyme-catalyzed reactions. *Trends Biochem. Sci.* **1982**, *7*, 102–105. [CrossRef]
27. Powers, J.C.; Asgian, J.L.; Ekici, O.D.; James, K.E. Irreversible inhibitors of serine, cysteine, and threonine proteases. *Chem. Rev.* **2002**, *102*, 4639–4750. [CrossRef]
28. Esteban, G.; Allan, J.; Samadi, A.; Mattevi, A.; Unzeta, M.; Marco-Contelles, J.; Binda, C.; Ramsay, R.R. Kinetic and structural analysis of the irreversible inhibition of human monoamine oxidases by ASS234, a multi-target compound designed for use in Alzheimer's disease. *Biochim. Biophys. Acta* **2014**, *1844*, 1104–1110. [CrossRef]
29. Ramsay, R.R.; Tipton, K.F. Assessment of enzyme inhibition: A review with examples from the development of monoamine oxidase and cholinesterase inhibitory drugs. *Molecules* **2017**, *22*, 1192. [CrossRef]
30. Dyshlovoy, S.A.; Hauschild, J.; Amann, K.; Tabakmakher, K.M.; Venz, S.; Walther, R.; Guzii, A.G.; Makarieva, T.N.; Shubina, L.K.; Fedorov, S.N.; et al. Marine alkaloid monanchocidin A overcomes drug resistance by induction of autophagy and lysosomal membrane permeabilization. *Oncotarget* **2015**, *6*, 17328–17341. [CrossRef]
31. Bakunina, I.Y.; Kol'tsova, E.A.; Pokhilo, N.D.; Shestak, O.P.; Yakubovskaya, A.Y.; Zvyagintseva, T.N.; Anufriev, V.F. Effect of 5-hydroxy- and 5,8-dihydroxy-1,4-naphthoquinones on the hydrolytic activity of alpha-galactosidase. *Chem. Nat. Comp.* **2009**, *45*, 69–73. [CrossRef]
32. Gote, M.M.; Khan, M.I.; Gokhale, D.V.; Bastawde, K.B.; Khire, J.M. Purification, characterization and substrate specificity of thermostable α-galactosidase from *Bacillus stearothermophilus* (NCIM-5146). *Process Biochem.* **2006**, *41*, 1311–1317. [CrossRef]
33. Borisova, A.S.; Reddy, S.K.; Ivanen, D.R.; Bobrov, K.S.; Eneyskaya, E.V.; Rychkov, G.N.; Sandgren, M.; Stålbrand, H.; Sinnott, M.L.; Kulminskaya, A.A.; et al. The method of integrated kinetics and its applicability to the exo-glycosidase-catalyzed hydrolysis of p-nitrophenyl glycosides. *Carbohydr. Res.* **2015**, *412*, 43–49. [CrossRef] [PubMed]
34. *MOE, 2018.01*; Chemical Computing Group ULC: 1010 Sherbrooke St. West, Suite #910, Montreal, QC, Canada, 2018.
35. Bakunina, I.Y.; Balabanova, L.A.; Golotin, V.A.; Slepchenko, L.A.; Isakov, V.V.; Rasskazov, V.V. Stereochemical course of hydrolytic reaction catalyzed by alpha-galactosidase from cold adaptable marine bacterium of genus *Pseudoalteromonas*. *Front. Chem./Chem. Biol.* **2014**, *2*, 1–6. [CrossRef] [PubMed]
36. Golotin, V.A.; Balabanova, L.A.; Noskova, Y.A.; Slepchenko, L.A.; Bakunina, I.Y.; Vorobieva, N.S.; Terentieva, N.A.; Rasskazov, V.A. Optimization of cold-adapted α-D-galactosidase expression in *Escherichia coli*. *Protein Expr. Purif.* **2016**, *123*, 14–18. [CrossRef] [PubMed]
37. Balabanova, L.A.; Golotin, V.A.; Bakunina, I.Y.; Slepchenko, L.V.; Isakov, V.V.; Podvolotskaya, A.B.; Rasskazov, V.A. Recombinant α-*N*-Acetylgalactosaminidase from marine bacterium modifying A erythrocyte antigens. *Acta Nat.* **2015**, *7*, 117–120.
38. Bradford, M.M. A rapid and sensitive method for the quantitation of microgram quantities of protein utilizing the principle of protein-dye binding. *Anal. Biochem.* **1976**, *72*, 248–254. [CrossRef]

39. Varfolomeev, S.D. *Khimicheskaya enzimologiya (Chemical Enzymology)*; Akademiya: Moscow, Russian, 2005.
40. Shared Resource Center Far Eastern Computing Resource of Institute of Automation and Control Processes Far Eastern Branch of the Russian Academy of Sciences (IACP FEB RAS). Available online: https://cc.dvo.ru (accessed on 22 June 2016).

Review

A Systematic Review of Recently Reported Marine Derived Natural Product Kinase Inhibitors

Te Li [1], Ning Wang [2], Ting Zhang [1], Bin Zhang [1,*], Thavarool P. Sajeevan [3], Valsamma Joseph [3], Lorene Armstrong [4], Shan He [1], Xiaojun Yan [1] and C. Benjamin Naman [1,5,*]

1 Li Dak Sum Yip Yio Chin Kenneth Li Marine Biopharmaceutical Research Center, Department of Marine Pharmacy, College of Food and Pharmaceutical Sciences, Ningbo University, Ningbo 315800, Zhejiang, China
2 Institute of Drug Discovery Technology, Ningbo University, Ningbo 315211, Zhejiang, China
3 National Centre for Aquatic Animal Health, Cochin University of Science and Technology, Fine Arts Avenue, Kochi, Kerala 682016, India
4 Department of Pharmaceutical Sciences, State University of Ponta Grossa, Ponta Grossa, Paraná 84030900, Brazil
5 Center for Marine Biotechnology and Biomedicine, Scripps Institution of Oceanography, University of California, San Diego, La Jolla, CA 92093, USA
* Correspondence: binzhang86@126.com (B.Z.); bnaman@nbu.edu.cn (C.B.N.); Tel.: +86-574-87600458 (B.Z.); +86-574-87604388 (C.B.N.)

Received: 25 July 2019; Accepted: 18 August 2019; Published: 23 August 2019

Abstract: Protein kinases are validated drug targets for a number of therapeutic areas, as kinase deregulation is known to play an essential role in many disease states. Many investigated protein kinase inhibitors are natural product small molecules or their derivatives. Many marine-derived natural products from various marine sources, such as bacteria and cyanobacteria, fungi, animals, algae, soft corals, sponges, etc. have been found to have potent kinase inhibitory activity, or desirable pharmacophores for further development. This review covers the new compounds reported from the beginning of 2014 through the middle of 2019 as having been isolated from marine organisms and having potential therapeutic applications due to kinase inhibitory and associated bioactivities. Moreover, some existing clinical drugs based on marine-derived natural product scaffolds are also discussed.

Keywords: kinase inhibitors; drug development; marine natural products

---

1. Introduction

Oceans and seas occupy almost 70% of the Earth's surface, and are estimated to host about 80% of all living species [1]. The rich biodiversity of the marine environment has been shown to yield equally rich chemical diversity of marine natural products isolated from the different organisms that have been studied since the 1940s [2]. In the same time period, roughly two-thirds of small molecules drugs with FDA or equivalent regulatory agency approvals have been granted for natural products or derivatives thereof [3]. This has certainly included many exceptionally impactful antibiotic or antiparasitic drugs and oncology therapeutics, among others [4–7]. Indeed, the published structures of natural products appear to correlate well with the chemical space occupied by approved drugs, and the natural product drug discovery continues to yield new and interesting compounds [2,8,9]. According to the MarinLit database (http://pubs.rsc.org/marinlit), more than 28,000 marine natural products have been reported after being isolated from a variety of marine sources; such as algae, ascidians, bryozoa, corals, microorganisms, sea hares, sea squirts, sponges, etc. [10]. The known and yet-to-be discovered marine natural product compounds represent a vast natural resource library. In the golden age of natural product research, many marine-derived scaffolds have been applied to clinical drug discovery and development due to new ideas and breakthroughs in screening technologies: (a) The mesophotic zone is increasingly recognized as being valuable for the discovery of new drug structures and unique activities [11]; (b) Computational methodologies play an essential role in the exploration of the biological activity and molecular structural networking of marine natural products [12]; (c) Databases of marine natural products (e.g. MarinLit: http://pubs.rsc.org/marinlit/ and Dictionary of Marine Natural Products: http://dmnp.chemnetbase.com/) are available to facilitate dereplication and discovery efforts and improve marine natural products research; (d) Chemical synthesis of complex molecules has become increasingly feasible at even gram-scale quantities that allow for the further biological interrogation and drug development of natural products previously isolated in low yields [13].

Most of the kinase inhibitor drugs approved thus far are adenosine triphosphate (ATP) competitive inhibitors that have various practical off-target liabilities. Four major classes of kinases exist in mammalian signaling pathways, and these can be classified broadly by substrate specificity: Serine/threonine kinases, tyrosine kinases, dual kinases (both Ser/Thr and Tyr), and lipid kinases [14]. Protein kinases share a common mechanism that is demonstrated in Figure 1, in which the protein function is activated or inactivated by the transfer of a phosphate group from ATP to the free hydroxy of a serine, threonine or tyrosine on the targeted protein, whereas protein phosphatases remove a phosphate group from phosphorylated amino acids and thereby effectively reverse the effect [15–18]. There are 38 FDA-approved small molecule kinase inhibitors available for the treatment of different diseases, such as cancers, immunological, inflammatory, degenerative, metabolic, cardiovascular and infectious diseases [19,20], and many more candidates still in clinical development. Some new biological therapeutics, in the way of monoclonal antibodies, have been discovered or engineered to successfully target some kinases [21,22]. A major challenge in the development of new kinase inhibitors is to overcome the toxicity or non-specificity of the existing drugs that act as ATP binding site competitors. Since small-molecule natural products are produced by and interact with proteins in their natural settings, some are known to function as signaling molecules among many forms of life and have been creatively repurposed for human health applications. Accordingly, some marine natural products have served as drug lead compounds, and these provide an abundant resource for the discovery of next-generation kinase inhibitors that target allosteric regions away from the ATP binding sites [23–26] or by stabilizing inactive conformations to prevent the function of certain kinases [27–30]. This is particularly relevant for the treatment of cancers and bacterial infections.

This review provides insight into the kinase inhibitors isolated from marine sources (bacteria and cyanobacteria, fungi, animals, algae, soft corals and sponges) that have been reported in the last five years since 2014 and highlight the associated biological activities and potential clinical applications. Examples of successful applications of marine natural products lead compound discovery and use of new drug screening technologies that have enabled rapid expansion in the field are accordingly provided.

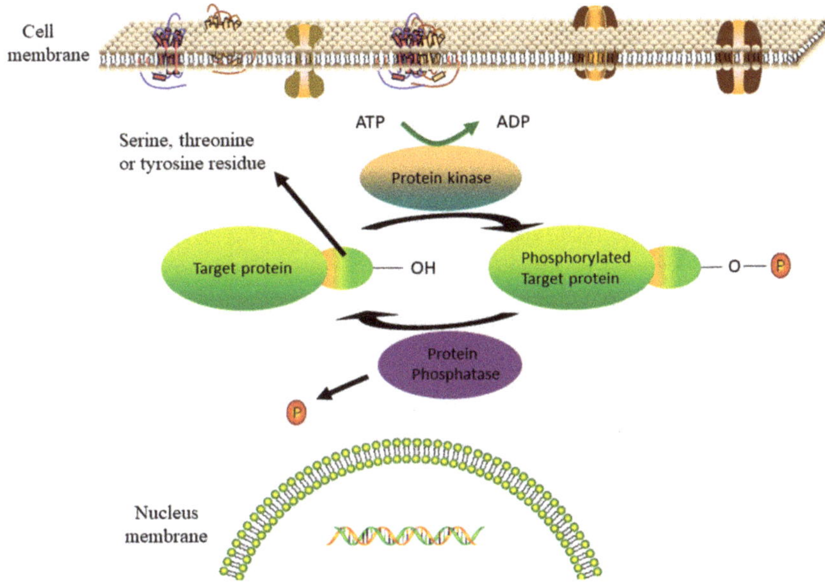

Figure 1. The catalytic cycle for protein phosphorylation by a protein kinase. Red circles represent phosphate groups. Adapted from content in Ref [16].

2. Discussion

2.1. New Kinase Inhibitors Discovered from Marine Organisms

Marine-derived natural products include a wide range of molecular classes such as alkaloids, macrolides, polypeptides, and terpenoids, and have different and interesting structures that play an important role in biological activities and clinical drug applications. As it pertains to the kinase research, a number of marine-derived kinase inhibitors have come from many different sources and target many different protein kinases (Figure 2). In the last five years covered in this review, from 2014–2019, there have been many results reported in the kinase research, and accordingly new biologically active kinase inhibitors that were discovered from various ocean life forms that include bacteria and cyanobacteria, fungi, animals, algae, soft corals and sponges. The initial reports of these natural product small molecules most typically include some rudimentary bioactivity test data, and further research can or does go on to explore more deeply the pharmacological potential of each. The new natural products reported during the period of this review are described systematically in the following sub-sections, divided by the source organisms reported for each.

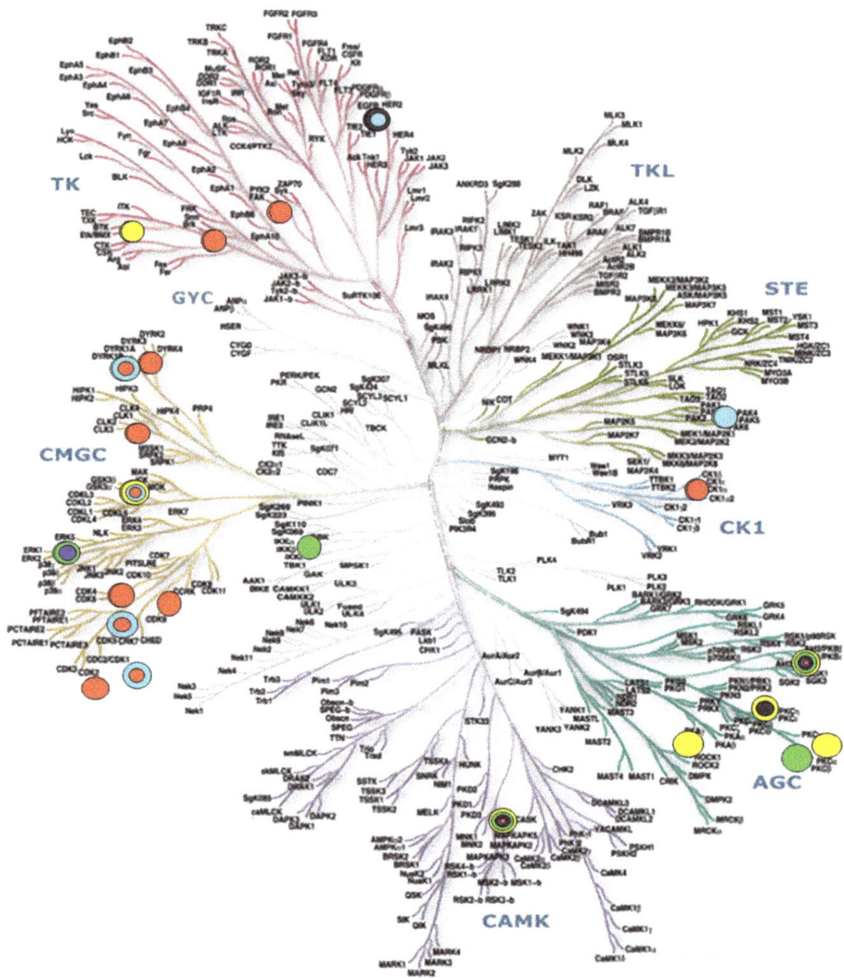

Figure 2. Availability of some marine-derived kinase inhibitors with activity on the phylogenetic tree of the human protein kinase family. Color codes indicate the producing or source organisms. Yellow: Marine bacteria. Green: Marine fungi. Gray: Marine soft coral. Light blue: Marine animals. Dark blue: Marine algae. Red: Marine sponges. Adapted from refs [31] and [32] with permission, the original illustration reproduced courtesy of Cell Signaling Technology, Inc. (www.cellsignal.com), 2019.

2.1.1. Kinase Inhibitors from Marine Bacteria

In the time period covered in this review, 22 new natural product kinase inhibitors (**1–22**) were reported after being isolated from marine bacteria. These are listed in Table 1 and shown by structure in Figures 3–9. The specific kinases reportedly involved in the biological activities of these compounds are described in the following paragraphs. Since staurosporine and its derivatives have been isolated and reported from marine bacteria, these are considered for the purpose of this review as being marine-derived natural products even though staurosporine itself and other analogues are also produced by terrestrial organisms. These are accordingly summarized in this section, for basic discoveries, and also in Section 2.2, for the preclinical and clinical candidates.

Table 1. A summary of kinase inhibitors isolated from marine bacteria, 2014–2019.

Compound	Chemical Class	Marine Source	Drug Targets/Inhibitory Activity	References
1	lipopeptide	*Bacillus megaterium*	p-Akt, p-MAPK, p-GSK-3β	[33]
2 3–5	indolocarbazole alkaloids	*Streptomyces* sp. A65	inactive analogue PKC, BTK (0.25–1.91 μM)	[34]
6–8	indolocabazole derivatives	*Streptomyces* sp. A68	PKC-α, BTK, ROCK2 (0.91–1.84 μM)	[35]
9–12	cyclizidine alkaloid	*Streptomyces* sp. HNA39	ROCK2 (7.0–42 μM)	[36–39]
13	Indolocarbazole alkaloid	*Streptomyces* sp. DT-A61	ROCK2 (5.7 nM)	[40]
14	indolocarbazole alkaloid	*Streptomyces* sp. DT-A61	PKC-α (92 nM)	[40]
15	indolocarbazole alkaloid	*Streptomyces* sp. A22	PKC, BTK, ROCK2 (0.91–1.84 μM)	[41]
16 17–21	staurosporine derivatives	*Streptomyces* sp. NB-A13	inactive analogue PKC-θ (0.06–9.43 μM)	[42]
22	naphthoquinone	*Streptomyces* sp. XMA39	ROCK2 (>20 μM)	[43]

In 2015, Goutam Dey et al. reported the anticancer activity of a lipopeptide, iturin A (**1**, Figure 3), which was isolated from the marine bacterium *Bacillus megaterium* [33]. In the same report, compound **1** was shown to inhibit p-Akt kinase, p-MAPK and p-GSK3β [33]. In other investigations, iturin A has been found to induce antiproliferative and apoptotic effects in breast cancer cells in vitro (IC_{50} values of MDA-MB-231, MCF-7,MDA-MB-468 and T47D cells were 7.98 ± 0.19, 12.16 ± 0.24, 13.30 ± 0.97 and 26.29 ± 0.78 μM, respectively) and in vivo [33]. These results suggest that iturin A, with its mechanism of Akt inhibition, may be useful for the development of a drug for this or other types of cancers. In 2018, Biao Zhou et al. reported the isolation of three new indolocarbazole alkaloids (**2–4**) along with nine known compounds (including **5**) from the marine bacterium *Streptomyces* sp. A65 [34]. Compounds **3–5** (Figure 4) inhibited PKC and BTK (IC_{50} between 0.25 μM and 1.91 μM), and compound **2** was not active [34]. The structural similarities of **2–5** along with the observed biological activities that include IC_{50} values across two or more orders of magnitude show the beginnings of a natural structural activity relationship (SAR). It would be interesting if this could be expanded by accessing further analogues through targeted isolation studies or synthetic chemistry.

In 2018, Le-Le Qin et al. reported two new indolocabazole analogues, compounds **6** and **7** (Figure 4) along with a related compound (**8**, Figure 5), isolated from a marine bacterium *Streptomyces* sp. A68 [35]. These compounds inhibited PKC-α, ROCK II and BTK with IC_{50} values ranging from 0.17 to 3.24 μM, and had cytotoxic activity in vitro against PC3 human prostate cancer cells [35]. The indolocarbazole alkaloid scaffold represented in these molecules has also been popular for the development of kinase inhibitory drugs. Comparing compounds **2–7**, it can be considered that the position of the oxygen atom on the tetrahydrofuran subunit and its para-hydroxy and carbamic acid moieties may have important effects on the kinase inhibitor activity of each molecule.

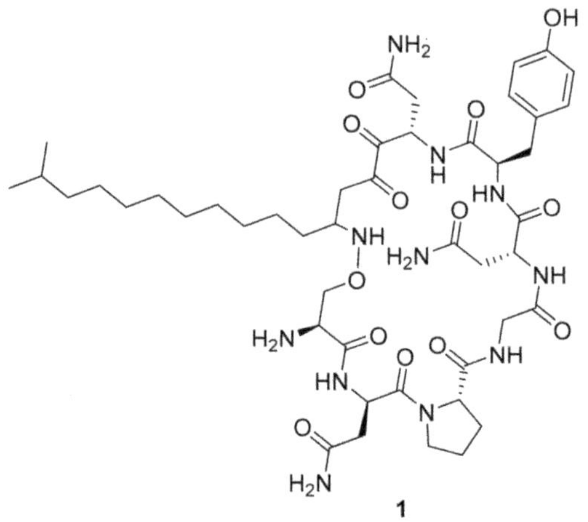

Figure 3. Structure of the lipopeptide iturin A (**1**).

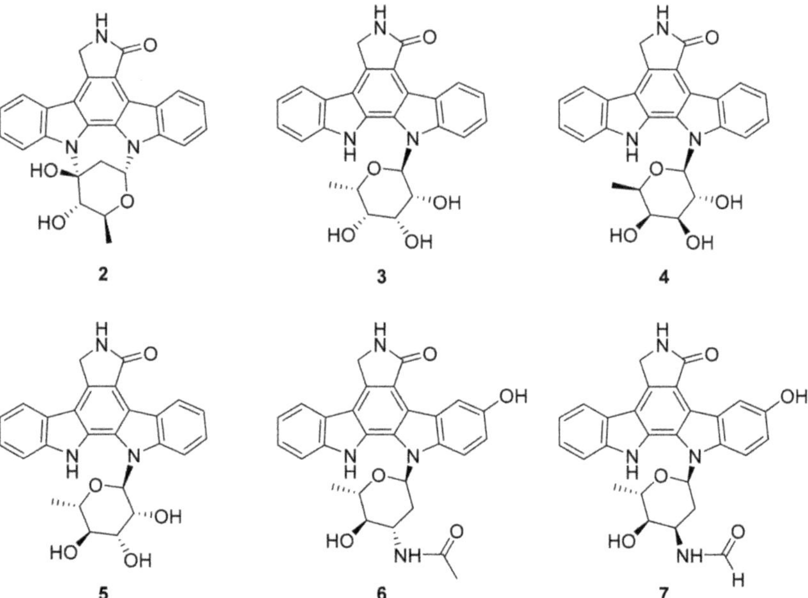

Figure 4. Structures of indolocarbazole alkaloids **2–7**.

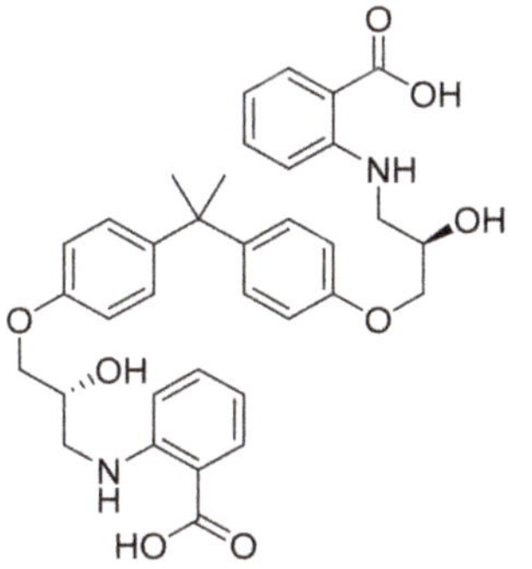

Figure 5. Structure of compound **8**.

In 2018, Yong-Jun Jiang et al. reported eight new and one known cyclizidine-type alkaloids including **9–12** (Figure 6) from the marine-derived actinomycete *Streptomyces* sp. HNA39 [36]. This class of molecules was first discovered only in 1982 from *Streptomyces* sp. NCIB 11649 [37–39]. From this group, compound **9** inhibited the in vitro survival of PC3 human prostate (IC_{50} = 0.52 ± 0.03 μM) and HCT116 human colorectal (IC_{50} = 8.3 ± 0.1 μM) cancer cells [36]. Compound **9** has four hydroxy groups on the indolizine core subunit, which may be important for producing the selectivity observed, since the mechanism of action is likely related to the Michael acceptor on the side chain. Furthermore, compounds **9–12** had some moderate inhibition against the ROCK2 (IC_{50} 7.0 ± 0.8 to 42 ± 3 μM) [36].

In 2018, Jia-Nan Wang et al. reported nine new indolocarbazole alkaloids from the marine bacterium *Streptomyces* sp. DT-A61 [40]. Among these, compound **13** (Figure 7) was shown to inhibit ROCK2 (IC_{50} = 5.7 nM) [40]. The results of biological activity tests showed compound **14** (Figure 7) to be cytotoxic to PC3 human prostate cancer cells (IC_{50} = 0.16 μM) [40]. In 2017, Xiang-Wei Cheng et al. isolated eight known and one new indolocarbazole alkaloid, 12-*N*-methyl-k252c (**15**, Figure 7), from the marine bacterium *Streptomyces* sp. A22 [41]. From this sample set, only **15** reportedly inhibited protein kinases (IC_{50} = 0.91–1.84 μM against various targets) [41].

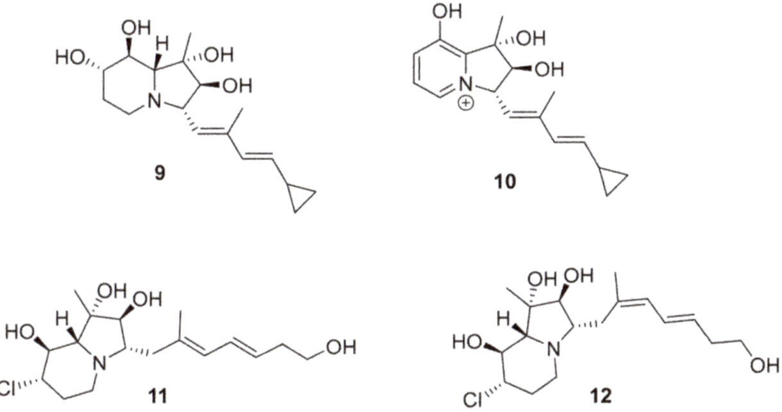

Figure 6. Structures of cyclizidine-type alkaloids **9–12**.

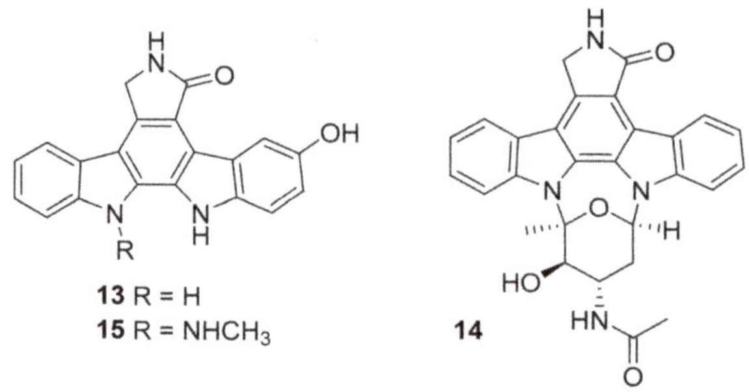

Figure 7. Structure of indolocarbazole alkaloids **13–15**.

In 2018, Biao Zhou et al. 15 staurosporine derivatives from the marine bacterium *Streptomyces* sp. NB-A13 [42]. Among these, six were new (**16–21**; Figure 8), and each one except for **16** inhibited PKC-θ (IC_{50} 0.06 to 9.43 μM) [42]. The most active of them was compound **21**, which also inhibited SW-620 cells in vitro (IC_{50} = 9.99 nM) at even greater potency than staurosporine (IC_{50} = 25.10 nM) [42]. Structurally, compound **21** bears an additional hydroxy group in the bis-indole ring system that is shared with staurosporine, indicating one possible avenue for medicinal chemistry optimization of other molecules in this chemical class.

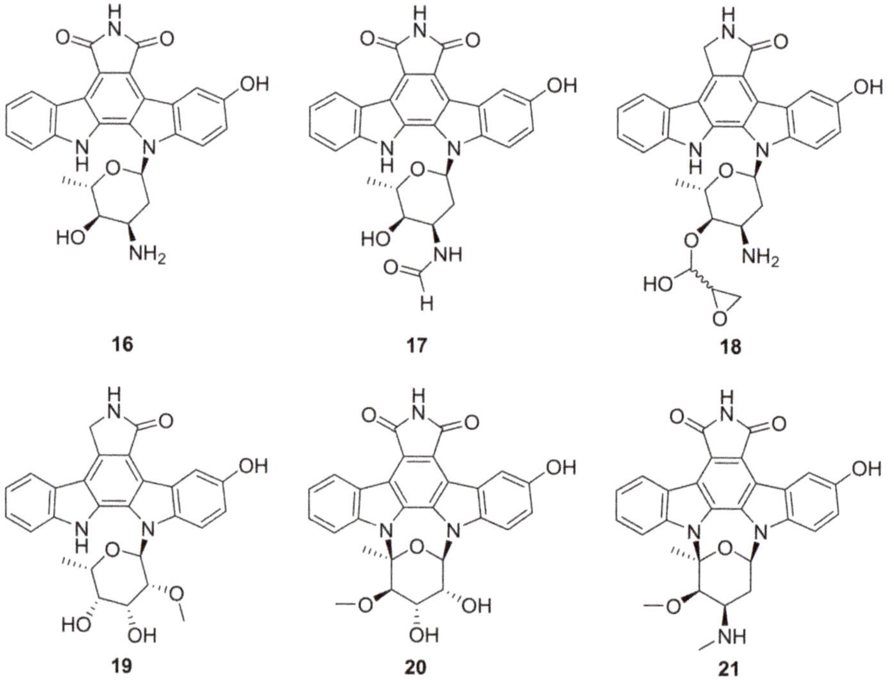

Figure 8. Structures of staurosporine derivatives **16–21**.

From the marine bacterium *Streptomyces* sp. XMA39, Yong-Jun Jiang et al. in 2018 isolated the known compound medermycin, and four new structurally related naphthoquinones, strepoxepinmycins A–D [43]. All of these marine natural products were growth inhibitors for different human pathogenic bacteria, and compound **22** (Figure 9) showed ROCK2 inhibition (IC_{50} > 20 μM) and cytotoxic activity against HCT-116 and PC3 cancer cells in vitro with IC_{50} > 40 mM [43].

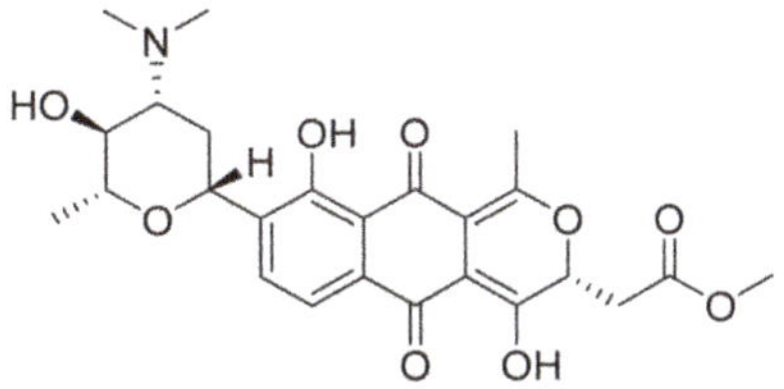

Figure 9. Structure of compound **22**.

2.1.2. Kinase Inhibitors from Marine Fungi

In the time period covered in this review, 16 new natural product kinase inhibitors (**23–38**) were reported after being isolated from marine fungi. These are listed in Table 2 and shown by structure in Figures 10–12. The specific kinases reportedly involved in the biological activities of these compounds are described in the following paragraphs.

In 2017, Dong-Ni Chen et al. reported the structure of sclerotiorin (**23**, Figure 10) isolated from the marine fungus *Penicillium* sp. Strain ZJ27 [44]. At the same time, sclerotiorin was shown to very weakly inhibit PknG (IC_{50} = 76.5 μM) [44]. In 2014, Dong-Cheol Kim et al. reported that methylpenicinoline (compound **24**, Figure 10) was isolated from marine fungus *Penicillium* sp. and inhibited p-p38 MAPK pathways, although no specific activity value was noted [45]. The authors proposed that **24** might be useful for further development toward the treatment of inflammatory and neuroinflammatory diseases, so the results of ongoing work may demonstrate their follow-up. Mycoepoxydiene (**25**, Figure 10) was reported in 2014 by Wen-Jiao Li et al., after it was isolated from the marine fungal strain *Diaporthe* sp. HLY-1 [46]. Compound **25** was shown to induce apoptosis and inhibit phosphorylation of IKK, and it was suggested that the inhibition of IKK activity and human cholangiocarcinoma (CCA) cells such as SK-ChA-1 (IC_{50} =12.5 ± 0.2 μM) and Mz-ChA-1 (IC_{50} = 35.6 ± 0.2 μM) may fully or partially explain the observed apoptosis [46]. These results suggest that compound **25** could be used as a lead molecule for the design and development of potent and selective new inhibitors of IKK with a potential for therapy of human cholangiocarcinoma disease [15,46]. In 2014, Kunlai Sun et al. reported a new indole-diterpenoid (**26**, Figure 10) that was isolated from the marine fungus *Aspergillus flavus* OUCMDZ-2205 [47]. This compound was found to inhibit PKC-β (IC_{50} = 15.6 μM), at 10 μM arrested A549 cells in the cell cycle S phase, and showed antibacterial activity against *Staphylococcus aureus* (MIC = 20.5 μM) [47]. In 2015, Wen-Liang Chen et al. isolated xyloketal B (compound **27**, Figure 10) from the fungus *Xylaria* sp. (No. 2508) that was collected in a mangrove [48]. Compound **27** was shown at 300 μM to inhibit p-Akt (inhibition rate ~ 34%) and p-ERK1/2 (inhibition rate ~ 40%) protein expression in vitro, and the determinants of anti-proliferation and migration effects against glioblastoma U251 cells (IC_{50} = 287.1 ± 1.0 μM) in vitro included TRPM7-regulated PI3K/Akt and MEK/ERK signaling [48].

Table 2. A summary of kinase inhibitors isolated from marine fungi, 2014–2019.

Compound	Chemical Class	Marine Source	Drug Targets/Inhibitory Activity	References
23	isochromen-6-one	*Penicillium* sp. ZJ27	PKnG (76.5 μM)	[41]
24	alkaloid	*Penicillium* sp.	p-p38 MAPK	[45]
25	polyketide	*Diaporthe* sp. HLY-1	IKK	[46]
26	indole diterpenoid	*Aspergillus flavus* OUCMDZ-2205.	PKC-β (15.6 μM)	[47]
27	furan derivative	*Xylaria* sp. (No. 2508)	p-Akt, p-ERK1/2	[48]
28	dihydro-isocoumarin	*Aspergillus* sp. SF-5976	p-p38 MAPK	[49]
29	phenylspiro-drimane derivative	*Stachybotrys* sp. KCB13F013	p-ERK, p-JNK, p-p38 MAPK	[50]
30			GSK-3β (0.35 ± 0.04 μM)	
31	benzocoumarin	*Aspergillus* sp. LF660	GSK-3β (0.13 ± 0.04 μM)	[51]
32			GSK-3β (0.20 ± 0.04 μM)	
33	lipophilic depsipeptide	*Alternaria* sp. SF-5016	p-JNK, p-p38 MAPK	[52]
34	citreohybridonol	*Toxicocladosporium* sp. SF-5699	p-p38 MAPK	[53]
35	2-methyl-hydroquinone	*Penicillium* sp. HL-85-ALS5-R004	p-Akt, p-ERK1/2	[54]
36	isopyrrolo-naphthoquinone	*Biscogniauxia mediterranea*	GSK-3β (8.04 μM)	[55]
37	polyketides	*Penicillium* sp. SF-5629	p-p38 MAPK	[56]
38		*Oidiodendron griseum* UBOCC-A-114129	CLK1 (15.6 g/mL)	[58]

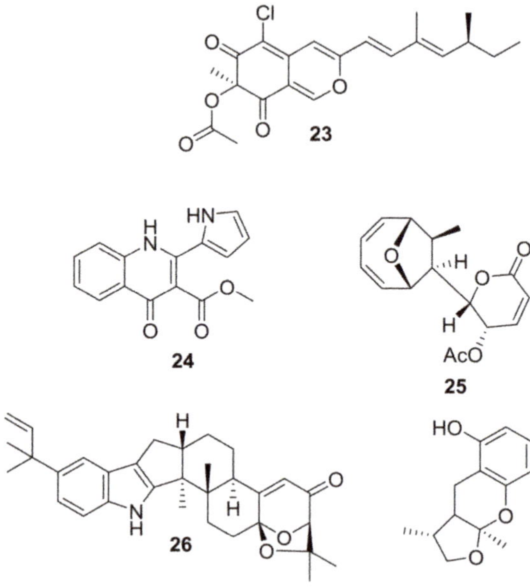

Figure 10. Structure of compounds **23**–**27**.

In 2015, Dong-Cheol Kim et al. [49] reported a new dihydroisocoumarin derivative (compound **28**, Figure 11) from the marine fungus *Aspergillus* sp. SF-5976, and this molecule inhibits p-p38 MAPK [49]. In 2016, Jong Won Kim et al. reported the discovery of stachybotrysin (**29**, Figure 11), which was isolated from the marine fungus *Stachybotrys* sp. KCB13F013 and found to inhibit the RANKL-induced activation of p-ERK, p-JNK and p-p38 MAPK [50]. The kinase inhibition activities of compound **28** and **29** were determined by Western blot analyses. In 2016, Jutta Wiese et al. reported pannorin (**30**), alternariol (**31**), and alternariol-9-methylether (**32**), which were isolated from the marine fungus, *Botryotinia fuckeliana* [51]. Compounds **30**–**32** (Figure 11) showed potent inhibition of GSK-3 in vitro with IC_{50} = 0.35 ± 0.04 μM, 0.13 ± 0.04 and 0.20 ± 0.04 μM, respectively [51]. The highly oxygenated benzocoumarin scaffolds represented in these molecules are apparently effective groups for GSK-3β inhibition, and these could merit further development through medicinal chemistry approaches although the current leads show very weak antibacterial and cytotoxic effects. In 2015, Wonmin Ko et al. reported alternaramide (**33**, Figure 11), which was isolated from the marine fungus *Alternaria* sp. SF-5016 [52]. This compound inhibited the formation of p-JNK and p-p38 MAPK, as determined by Western blotting, suggesting that it could be useful in the treatment of various acute, systemic, and neurological inflammatory diseases [52]. In 2015, Kwang-Ho Cho et al. reported citreohybridonol (**34**, Figure 11) after it had been isolated from the extract of the marine-derived fungus *Toxicocladosporium* sp. SF-5699 [53]. Compound **34** was pursuantly shown to inhibit the activation of p38 MAPK pathways by the Western blot analysis [53].

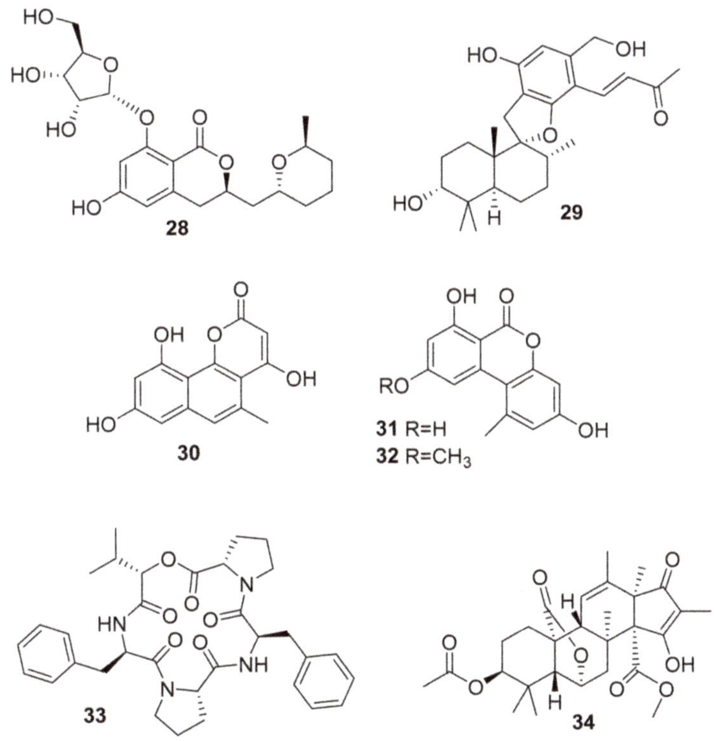

Figure 11. Structure of compounds **28–34**.

In 2016, M García-Caballero et al. reported that toluquinol (**35**, Figure 12), isolated from the marine fungus *Penicillium* sp. HL-85-ALS5-R004, inhibited the phosphorylation of p-Akt and p-ERK1/2 in vitro and VEGF-C-induced lymphatic vessel formation and corneal neovascularization in mice [54]. Compound **35** plays a demonstrated role in inhibiting angiogenesis and lymphangiogenesis, which is meaningful in connection with the implications of tumour-induced lymphangiogenesis and lymphatic metastasis. In 2016, Bin Wu et al. reported on biscogniauxone (**36**, 12), a new isopyrrolonaphthoquinone isolated from the deep-sea (2800 m) fungus *Biscogniauxia mediterranea* [55]. Compound **36** moderately inhibited GSK-3β (IC_{50} = 8.04 μM) and had almost negligible antibacterial activity against *Staphylococcus epidermidis* and methicillin-resistant *S. aureus* (IC_{50} ~ 100 μM) [55]. Marine-derived fungi from the deep sea are promising but severely under-studied biological resources, and these possess abundant novel and bioactive secondary metabolites that urgently need to be mined if adequate access to the resource can be obtained [57]. In 2017, Nguyen Thi Thanh Ngan et al. reported that citrinin H1 (compound **37**, Figure 12), isolated from the marine fungal strain *Penicillium* sp. SF-5629, inhibited p-p38 MAPK expression in vitro [56]. Additionally in 2017, Marion Navarri et al. reported that dihydrosecofuscin (compound **38**, Figure 12), isolated from a deep-sea (765 m) fungus *Oidiodendron griseum* UBOCC-A-114129, inhibited CLK1 (IC_{50} = 15.6 μg/mL) [58]. Compound **38** showed weak antibacterial activities (MIC ~ 100 μg/mL) against Gram-positive bacteria, and the authors noted a potential use of this CLK1 inhibitor for the treatment of Alzheimer's due to the connection between the particular kinase and disease [58].

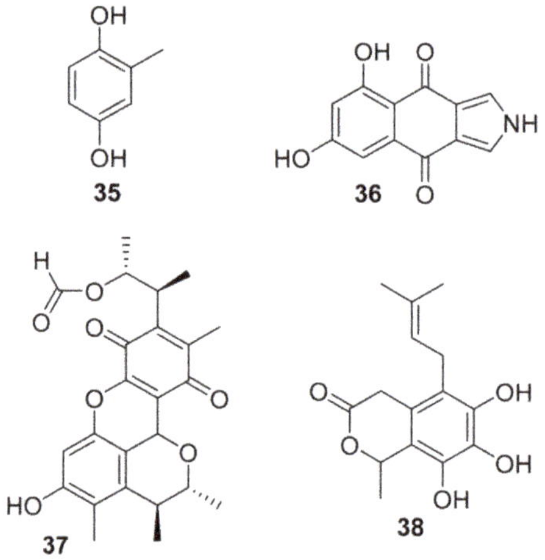

Figure 12. Structure of compounds **35**–**38**.

2.1.3. Kinase Inhibitors from Marine Soft Coral

In the time period covered in this review, two new natural product kinase inhibitors (**39** and **40**) were reported after being isolated from marine soft coral (see Table 3). Dihydroaustrasulfone alcohol (**39**, Figure 13), is a molecule that was reported in 2015 by Pei-Chuan Li et al. after being isolated from soft corals [59]. This compound inhibited ERK/MAPK and PI3K/AKT, and was suggested as a lead that could be pursued for the prevention and treatment of arterial restenosis [59]. In 2017, pachycladin A (**40**, Figure 13) was re-isolated from the soft coral *Cladiella pachyclados* found in the Red Sea, and a newly shown activity (absent from the previous report of this compound in 2010 [61]) was that it inhibited two members of the EGFR family when screened at 10 μM [60].

Table 3. A summary of kinase inhibitors isolated from marine soft coral, 2014–2019.

Compound	Chemical Class	Marine Source	Drug Targets/Inhibitory Activity	References
39	simple sulfone alcohol	*Cladiella australis*	MAPK, AKt	[59]
40	eunicellin diterpenoid	*Cladiella pachyclados*	EGFR	[60]

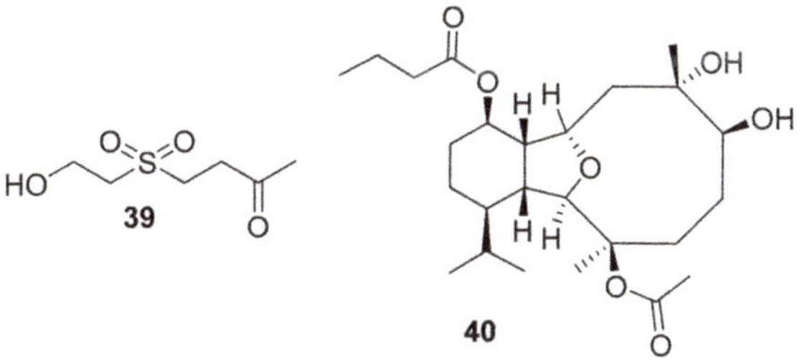

Figure 13. Structure of compounds **39–40**.

2.1.4. Kinase Inhibitors from Marine Animals

In the time period covered in this review, 14 new natural product kinase inhibitors (**41–54**) were reported after being isolated from marine animals. These are listed in Table 4 and shown by structure in Figures 14–16. The specific kinases reportedly involved in the biological activities of these compounds are described in the following paragraphs.

In 2014, didemnaketals D and E (compounds **41** and **42**, Figure 14) were reportedly discovered from marine tunicates of the genus *Didemnum*, and these compounds inhibited several kinases (CDK5, CK1, DyrK1A, and GSK3) at 10 μg/mL and were moderately antimicrobial in vitro against *S. aureus* and *Bacillus subtilis* [62]. In 2015, Youssef [63] et al. reported the three new purine alkaloids **43–46** (Figure 15), together with seven known compounds (**45–52**, Figure 15) from the marine tunicate *Symplegma rubra*, collected from the Saudi Red Sea coast (5–7 m depth). The authors reported that these compounds were found to be moderate inhibitors of different kinases (CDK5, CK1, DyrK1A, and GSK-3), with some activity at 10 μg/mL, but noted that the observed IC_{50} values (>10 μM) were "not sufficient to pursue with these compounds into the in vivo studies" [63]. This report was somewhat vague and did not specify the individual activity of each molecule.

Table 4. A summary of kinase inhibitors isolated from marine animals, 2014–2019.

Compound	Chemical Class	Marine Source	Drug Targets/Inhibitory Activity	References
41–42	spiroketals	marine ascidian genus *Didemnum*	CDK5, CK1, DyrK1A, GSK-3 (10 μg/mL)	[62]
43–52	purine alkaloids	marine tunicate *Symplegma rubra*	CDK5, CK1, DyrK1A, GSK-3 (10 μg/mL)	[63]
53	saponin sulfate	sea cucumber	PAK1 (1.2 μM) LIMK (60 μM) AKT (59 μM)	[64,65]
54	anthraquinone derivative	marine echinoderm *Comanthus* sp.	IGF1-R (5 μM) FAK (8.4 μM) EGFR (4 μM)	[66]

In 2017, frondoside A (compound **53**, Figure 16) was isolated from *Cucumaria frondosa*, the Atlantic sea cucumber, and reported as being an effective inhibitor of PAK1 (IC_{50} ~ 1.2 μM), LIMK (IC_{50} ~ 60 μM), AKT (IC_{50} ~ 59 μM) and A549 lung cancer cells (IC_{50} ~ 1–3 μM) in vitro, indicating that it may be useful in the treatment of malignancies [64,65]. Additionally in 2017, the molecule 1-deoxyrhodoptilometrin (compound **54**, Figure 16), which was first discovered by Wright et al. from the Echinoderm *Colobometra perspinosa* [67], was re-isolated from the marine echinoderm *Comanthus* sp. [66]. This compound was subsequently reported to inhibit the IGF1-R kinase, focal adhesion kinase, and EGF receptor kinase with IC_{50} = 5, 8.4 and 4 μM, respectively [66].

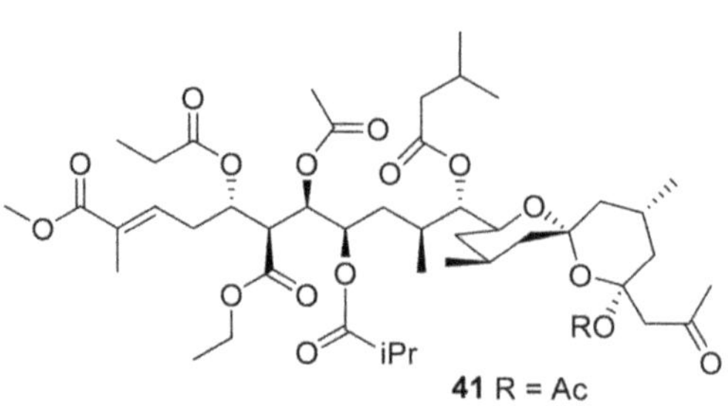

41 R = Ac
42 R = H

Figure 14. Structures of didemnaketals D (**41**) and E (**42**).

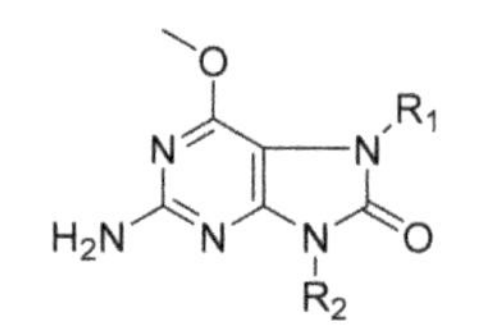

43 $R_1 = R_2 = CH_3$
44 $R_1 = H$ $R_2 = CH_3$
45 $R_1 = CH_3$ $R_2 = H$

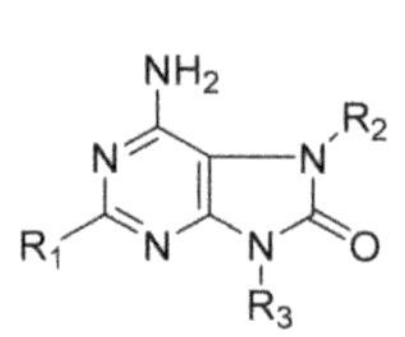

46 $R_1 = OCH_3$ $R_2 = CH_3$ $R_3 = H$
47 $R_1 = R_2 = H$ $R_3 = CH_3$
48 $R_1 = R_3 = H$ $R_2 = CH_3$
49 $R_1 = R_2 = R_3 = H$

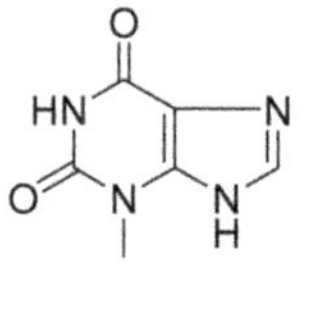

50

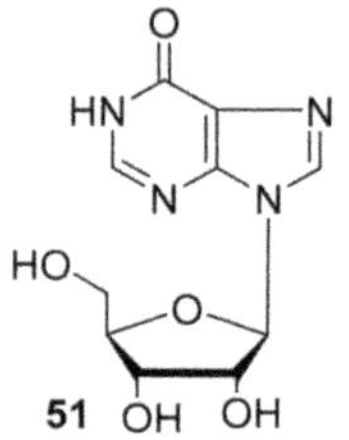

51

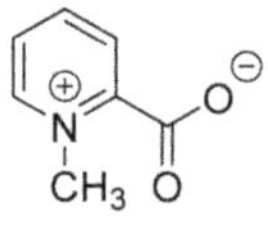

52

Figure 15. Structures of purine and pyridine alkaloids (**43**–**52**).

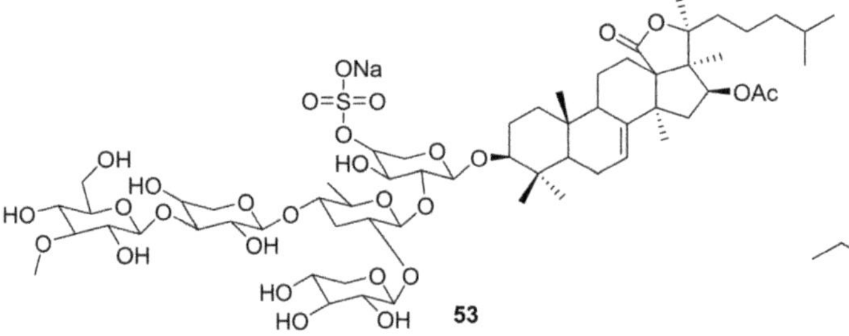

53

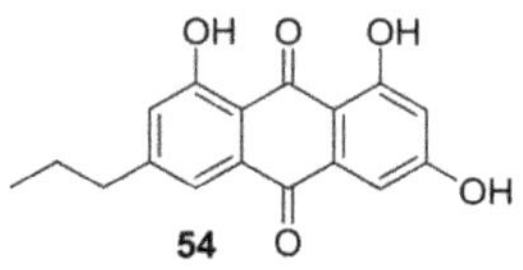

54

Figure 16. Structure of compounds **53**–**54**.

2.1.5. Kinase Inhibitors from Marine Algae

In the time period covered in this review, 10 new natural product kinase inhibitors (**55–64**) were reported after being isolated from marine algae. These are listed in Table 5 and shown by structure in Figure 17, Figure 18 and Scheme 1. The specific kinases reportedly involved in the biological activities of these compounds are described in the following paragraphs.

In 2015, Shuai-Yu Wang et al. reported compound **55** (Figure 17) as a multi-target RTKs inhibitor such as FGFR2, FGFR3, VEGFR2 and PDGFR-α (with in vitro inhibition rates of 57.7%, 78.6%, 78.5% and 71.1%, respectively), which was isolated from *Rhodomela confervoides*, a red alga, and this molecule also inhibits p-PKB/Akt [68,69]. Compound **55** may accordingly present a scaffold from which to develop new multi-target RTKs inhibitors. In 2017, Sung-Hwan Eom et al. reported that the polyphenolic molecule eckol (**56**, Figure 17), purified from the edible brown seaweed *Eisenia bicyclis*, inhibited *Propionibacterium acnes* mediated phosphorylation of Akt [70]. Eckol is also proposed as an anti-inflammatory agent and antioxidant. In 2017, fucoxanthin (**57**, Figure 17), a well-known marine carotenoid, was shown to inhibit Akt/NF-κB and involved MAPK pathways in vitro, suggesting that it could be a potential therapeutic agent acting by this mechanism for the therapy of neurodegenerative diseases [71–73]. Further validation studies are obviously necessary, and likely ongoing. In 2018, Youn Kyung Choi et al. reported the isolation of bis (3-bromo-4,5-dihydroxybenzyl) ether (**58**, Figure 17) from *Polysiphonia morrowii*, a red alga [74]. Compound **58** selectively inhibited p-ERK, demonstrating that it could have potential in the treatment of inflammatory diseases [74]. In 2014, Paudel et al. postulated that since the polyphenolics 6–6 bieckol and pholorofucofuroeckol A (**59** and **60**, Figure 17) inhibited p-ERK1/2 and p-JNK activation, these hold potential value for treating pulpitis and oral diseases [75]. These two compounds were isolated from *Eisenia bicyclis*, a brown alga [75].

Table 5. A summary of kinase inhibitors isolated from marine algae, 2014–2019.

Compound	Chemical Class	Marine Source	Drug Targets/Inhibitory Activity	References
55	bromophenol	red alga *Rhodomelaceae confervoides*.	RTKs, p-PKB/Akt	[68,69]
56	phlorotannin	brown seaweed *Eisenia bicyclis*	p-AKt	[70]
57	carotenoid	brown alga wakame seaweeds wakame (*Undaria pinnatifida*) and kombu (*Laminaria japonica*)	AKt, MAPK	[71–73]
58	bromophenol	red alga *Polysiphonia morrowii*	p-ERK	[74]
59–60	lipopoly-saccharide	brown seaweed marine alga (*Eisenia bicyclis*)	p-ERK1/2, p-JNK	[75]
61[a]	cyclic depsipeptide	marine cyanobacterium *Moorea producens* (formerly *Lyngbya majuscula*)	RTKs	[76]
62 [a]	bromo-honaucin	marine cyanobacterium *Leptolyngbya crossbyana*	AKt, ERK	[77,78]
63 [b]	cyclic depsipeptide	marine cyanobacterium *Moorea producens*	JNK	[79,80]
64	cyclic depsipeptide	marine cyanobacterium *Leptolyngbya* sp.	VEGFA/VEGFR2	[81,82]

[a] This molecule is synthetic but designed based on a marine natural product. [b] Hoiamide A is JNK activator.

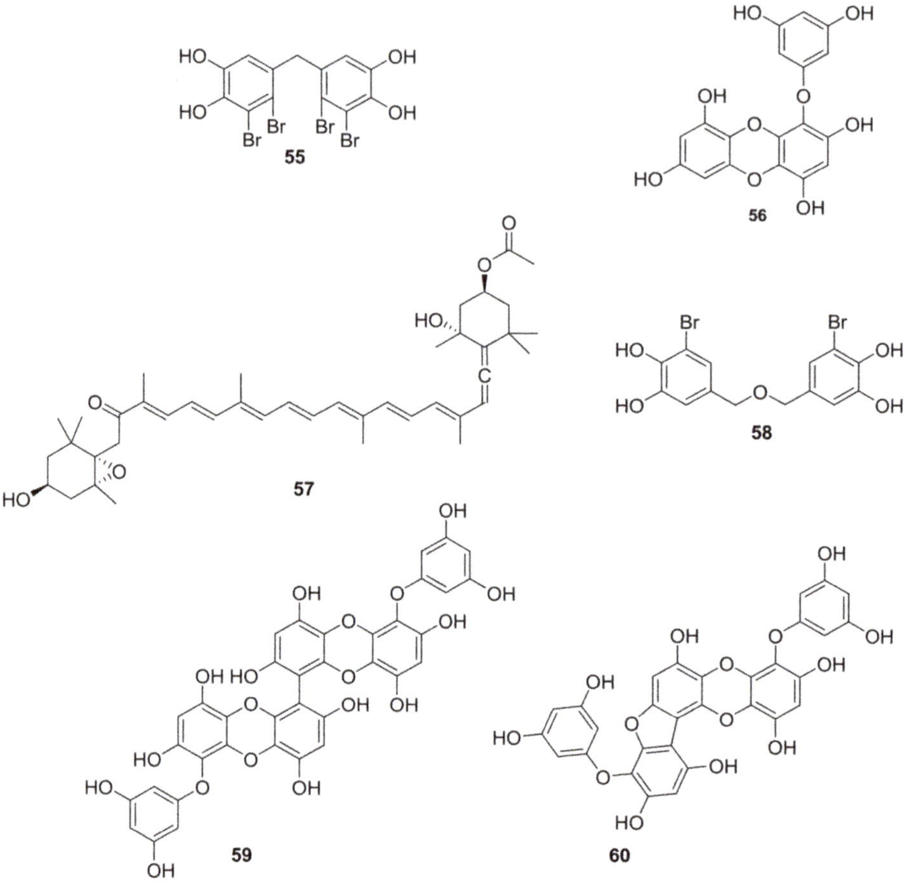

Figure 17. Structure of compounds **55–60**.

Apratoxin A is a cyanobacterial, or "blue-green algal", natural product macrocycle that was first reported by Luesch et al. in 2001 as a product of a hybrid polyketide synthase-nonribosomal peptide synthetase (PKS/NRPS) biosynthetic pathway [83]. In 2017, Wei-Jing Cai et al. later designed and synthesized a series of cyclic depsipeptides based on apratoxin A, including apratoxin S10 (**61**, Scheme 1) [76]. This analogue of apratoxin A was selected as a desirable drug lead with cancer cell growth (IC_{50} = 5.97 nM against HCT116) and RTKs inhibitory activity [76]. The synthetic route of generating apratoxin S10 is depicted in Scheme 1. The synthesis of apratoxin S10 followed the reported strategy used to produce the lead compound, apratoxin A, with some modification to introduce new functionality [76].

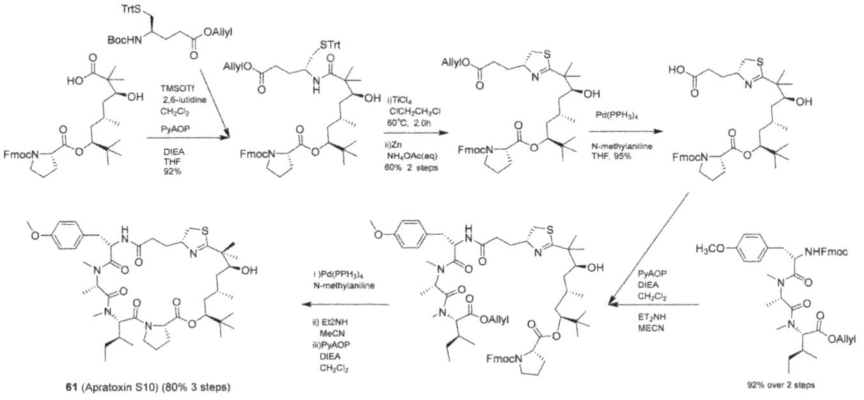

Scheme 1. Synthesis and structure of apratoxin S10 (**61**).

In 2015, Mahesh Sapkota [78] et al. reported bromo-honaucin A (**62**, Figure 18), which was first isolated from marine cyanobacterium *Leptolyngbya crossbyana* and chemical synthesis was completed by Hyukjae Choi and William H. Gerwick et al. [77]. Compound **62** inhibited AKt and ERK, and the authors noted a potential use of this AKt and ERK inhibitor for the treatment of bone lysis due to the connection between the particular kinase and disease. In 2015, Zhengyu Cao et al. [81] reported the discovery of hoiamide A (**63**, Figure 18), which was first isolated from marine cyanobacterium *Moorea producens* by Alban Pereira et al. [79] and found to stimulate and increase the hoiamide A-induced activation of JNK. Compound **63** showed the inhibition of neurite outgrowth (IC_{50} = 4.89 nM). In 2016, Jeffrey D. Serrill et al. [82] reported coibamide A (**64**, Figure 18) (first isolated from marine cyanobacterium *Leptolyngbya* sp. [81]) inhibited VEGFA/VEGFR2 expression in mice and showed antitumor activities against glioblastoma xenografts.

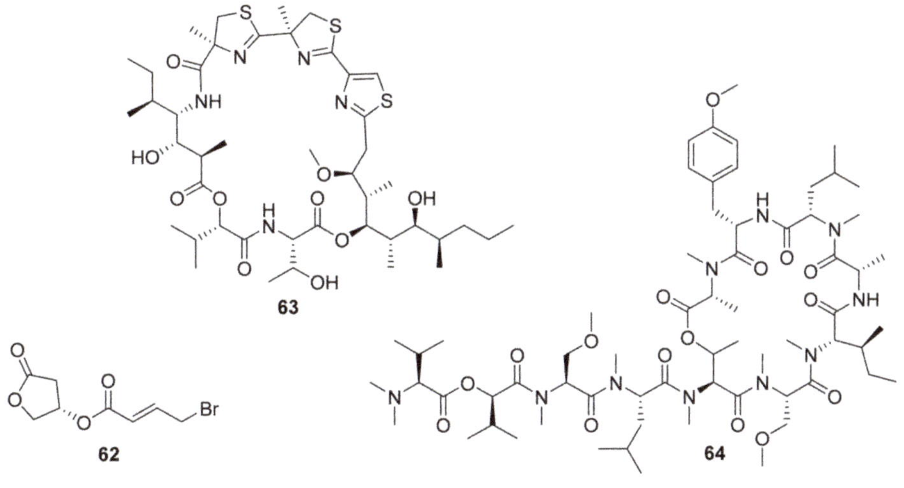

Figure 18. Structure of compounds **62**–**64**.

2.1.6. Kinase Inhibitors from Marine Sponges

In the time period covered in this review, 42 new natural product kinase inhibitors (**65**–**106**) were reported after being isolated from marine sponges. These are listed in Table 6, shown by structure in Figures 19–26 and Schemes 2–5, and described in detail in the following paragraphs.

In 2014, Fabien Plisson et al. reported some new bromopyrrole alkaloids including hymenialdisine (**65**, Figure 19) that were isolated from *Callyspongia* sp. [84]. This interesting class of molecules has been studied for several decades, owing to their peculiar structural skeleton and associated pharmacological effects [85,86]. Compound **65** inhibited CK1, CDK5, GSK3β and SW620 (3.1 μM), KB-3-1 (2.0 μM), which are targets for treating neurodegenerative diseases [84]. In 2014, the β-carboline alkaloids **66–68** (Figure 19), including hainanerectamines B and C, were isolated from the marine sponge *Hyrtios erecta* after it was collected in Hainan, China [87]. These molecules have been shown to inhibit the dual kinase Aurora A with relatively high IC_{50} values of 10 μg/mL–25 μg/mL [88]. In 2014, Mohamed R. Akl et al. reported the isolation of sipholenol A from the sponge *Callyspongia siphonella*, collected in the Red Sea [89]. A semisynthetic analogue of sipholenol A, sipholenol A-4-*O*-3′,4′-dichlorobenzoate (**69**, Figure 19) was associated with the suppression of the Brk and FAK signaling pathway in vitro and in vivo, also inhibited MDA-MB-231, MCF-7, BT-474 and T-47D breast cancer cells with IC_{50} values of 7.5 μM, 15.2 μM, 20.1 μM and 25.1 μM, respectively, making it a potentially interesting pro-drug for sipholenol A [89].

Table 6. A summary of kinase inhibitors isolated from marine sponges, 2014–2019.

Compound	Chemical Class	Marine Source	Drug Targets/Inhibitory Activity	References
65	bromopyrrole alkaloid	*Callyspongia* sp. (CMB-01152).	CK1, CK5, GSK3β	[84]
66 **67** **68**	indole alkaloids	*Hyrtios erecta*	STK (24.5 μg/mL) STK (13.6 μg/mL) STK (18.6 μg/mL)	[88]
69	sipholane triterpenoid	*Callyspongia siphonella.*	p-Brk, p-FAK	[89]
70	indole alkaloid	*Geodia barretti*	RIPK2 (8.0 μM) CAMK1a (5.7 μM) SIK2 (6.1 μM)	[90,92,93]
71	diarylpyrazine	*Hamacanthins*	PDGF-Rβ (0.02 μM)	[93]
72 **73** [a]	tryptoline analogues	*Fascaplysinopsis* sp.	CDK4/D1 (0.35 μM) CDK4 (10 μM)	[94–98]
74 [a]	alkaloid derivative	*Fascaplysinopsis* sp.	PI3K/Akt	[99]
75 **76** **77**	bisindole alkaloids	*Topsentia pachastrelloides*	MRSA PK (60 nM) MRSA PK (16 nM) MRSA PK (1.4 nM)	[100]
78	macrocyclic oxaquinolizidine alkaloid	*Xestospongia* species.	PI3K, Akt	[101]
79–84	bromopyrrole alkaloids	*Stylissa massa* and *Stylissa flabelliformis*	GSK-3, DYRK1A, CK-1 (0.6–6.4 μM)	[102]
85 **86** **87–88**	adociaquinone derivatives	*Xestospongia* sp.	inactive analogue CDK9/cyclin T (3 μM) CDK5/p25 (6 μM)	[108]
89	cyclic peptide	*Amphibleptula*	GSK-3β	[109]
90	polycyclic quinone	*Petrosia*	hexokinase II	[110]
91	halogenated alkaloid	*Acanthostrongylophora ingens*	CK1δ/ε (6 μM)	[111]
92	isomalabaricane triterpenoid	*Jaspis stellifera*	PI3K, Akt	[112]
93–95	guanidine alkaloids	*Smenospongia* sp.	CDK2/6	[113]
96–97	2-aminoimidazolone alkaloids	*Leucetta* and *Clathrina*	DYRKs (0.17 ~ 0.88 μM) CLKs (0.32 ~ 8.6 μM)	[114]
98–103	manzamine alkaloids	*Acanthostrongylophora* sp.	MtSK	[115,116]
104	bis-indole alkaloid	*Topsentia pach astrelloides*	MRSA PK	[117]
105	sphingolipid	*Pachastrissa* sp.	SphK1 (7.5 μM) SphK1 (20.1 μM)	[118,119]
106	polyfused-benzofuran	*Aka coralliphaga*	PIK-α (100 nM)	[122]

[a] This molecule is synthetic but designed based on a marine natural product.

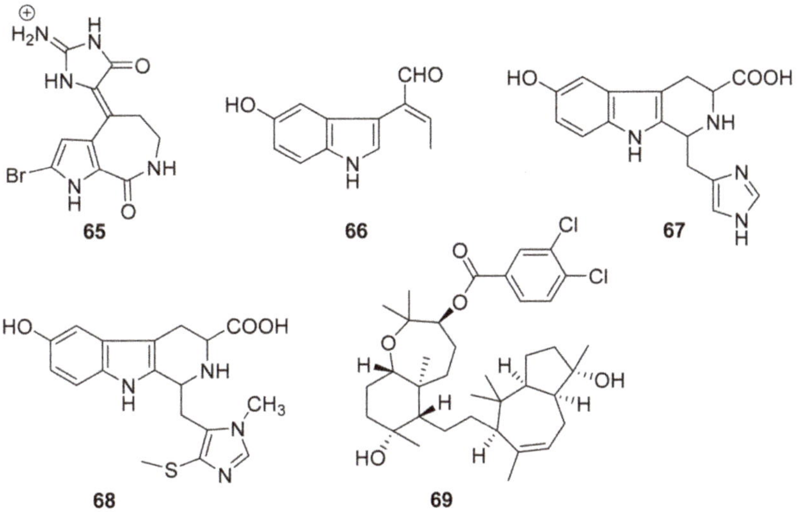

Figure 19. Structure of compounds **65–69**.

In 2015, Karianne F. Lind et al. reported that barettin (**70**, Figure 20), which was first isolated and described in 1986 [90], inhibited RIPK2 (IC_{50} = 8.0 μM), CAMK1a (IC_{50} = 5.7 μM), SIK2 (IC_{50} = 6.1 μM) and produced anti-inflammatory effects in vitro [91,92]. In 2015, Rebecca Horbert et al. reported the synthetic compound **71** (Figure 20) with a 3,5-diarylpyrazin-2-one core that was based on hamacanthins A and B, which are sponge natural products of the bis-indole alkaloid class [93]. Compound **71** inhibited PDGF-Rβ with good potency (IC_{50} = 0.02 μM), and was cytotoxic in vitro against PDGF-R dependent cancer cells [93]. The accompanying structure-activity relationship indicated that the added 3′-methoxy and 4′-hydroxyphenyl substituents led to an increase in the inhibitory activity against PDGF-Rβ, helping to direct future related research studies [93]. Fascaplysin (**72**, Figure 20) previously was isolated from the marine sponge *Fascaplysinopsis* species [94] and shown to inhibit CDK-4/D1 with moderate potency (IC_{50} = 0.35 μM) and some selectivity amongst kinases [95]. This compound is known to have multiple other mechanisms of action, including DNA binding. More recently, in 2015, S. Mahale et al. synthesized a non-planar analogue of fascaplysin, *N*-(biphenyl-2-yl) tryptoline (**73**, Figure 20), and reported that this retained some weak but selective inhibition of CDK-4/D1 (IC_{50} ~10 μM) [96]. Although this represents a marked drop in potency, compound **73** was successfully designed to overcome DNA intercalation that had been considered a liability of the more active **72** [96]. Compound **73** was validated as being a preferable anticancer agent than **72**, including the results from PK studies in mice [96]. It has been further suggested that compound **73** constitutes a lead scaffold from which drug candidates could be designed for the treatment of certain cancers [96–98]. The synthetic route of production for **73** is depicted in Scheme 2 [96], which includes commercially available starting materials (biphenyl 2-carboxylic acid and tetrahydro-β-carboline), commonly used reagents [hexafluorophosphate benzotriazole tetramethyl uronium (HBTU) and *N*,*N*-diisopropylethylamine (DIEA)], and runs in good yield (~70%). The further development of fascaplysin analogues and derivatives through medicinal chemistry will accordingly help overcome the common "supply problem" of advancement of natural product drug candidates. In 2017, Sonia Sharma et al. synthesized 4-chlorofascaplysin (**74**, Scheme 3; a marine sponge alkaloid fascaplysin analogue) and found it to be cytotoxic to breast cancer cells in vitro (IC_{50} = 0.3 μM) and to inhibit PI3K/AKt/mTOR in vitro and in vivo [99]. The synthetic route of 4-chloro fascaplysin is depicted in Scheme 3, where the two-step synthesis includes low cost materials and high yield reactions [99].

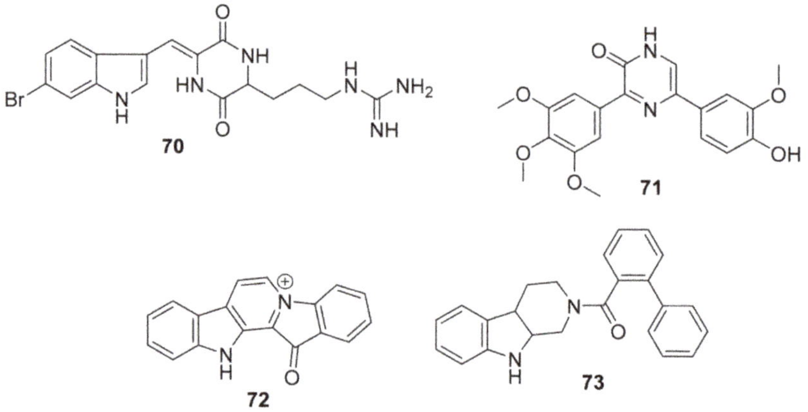

Figure 20. Structure of compounds **70**–**73**.

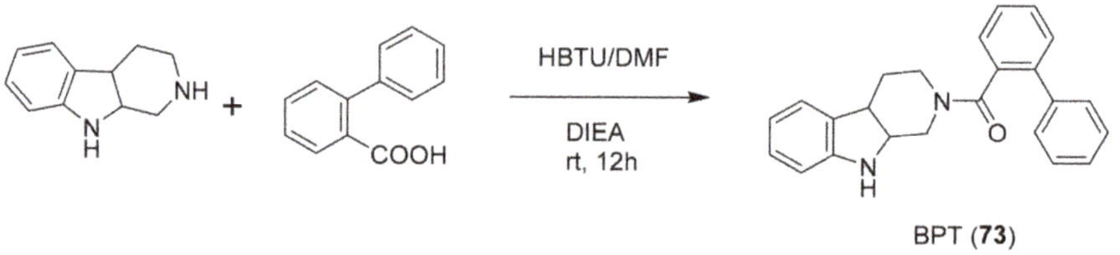

Scheme 2. Synthetic preparation of the fascaplysin derivative, compound **73**.

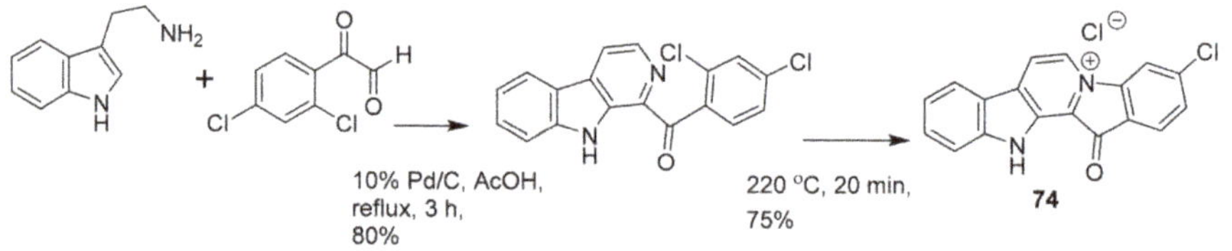

Scheme 3. Synthesis of 4-chloro fascaplysin (**74**).

In 2015, Clinton G. L. Veale et al. created a series of synthetic bisindole compounds including **75** and **76** (Figure 21), based on deoxytopsentin that was isolated from marine sponge *Topsentia pachastrelloides* [100]. Compounds **75** and **76** inhibit PK from methicillin resistant *S. aureus* (IC_{50} = 60 and 16 nM, respectively), and had good in vitro antibacterial activity (MIC ~ 10 μg/mL) [100]. Compound **77** (Figure 21) is among the series of synthetic analogues based on **75** and **76**, and it had enhanced inhibitory activity against MRSA PK (IC_{50} = 1.4 nM) [100]. Additionally in 2015, Mohamed R. Akl et al. reported an RTK inhibitor, araguspongine C (compound **78**, Figure 22), was isolated from the marine sponge *Xestospongia* sp. [101]. Sherif S. Ebada et al. reported the productive discovery of 25 bromopyrrole alkaloids in 2015, from the Indonesian marine sponges *Stylissa massa* and *Stylissa flabelliformis* [102]. This sample set included dispacamide E (compound **79**) [103], the brominated aldisines **80**–**82** [103,104], (-)-mukanadin C (**83**) [105,106] and (-)-longamide (**84**) [107], these compounds, shown in Figure 22, were moderate inhibitors of GSK-3, DYRK1A and CK-1 (IC_{50} 0.6 ~ 6.4 μM), and may have future development potential for treating various diseases [102]. In 2015, Fei He et al. reported seven new derivatives of adociaquinone that were isolated from a marine sponge, *Xestospongia* sp., collected in Indonesia [108]. Among these, compounds **85** and **86** (Figure 22) moderately but selectively inhibited CDK9/cyclin T and CDK5/p25 at 3–6 μM IC_{50}, while **87** and **88** (also Figure 22) moderately inhibited most protein kinases without selectivity (IC_{50} = 0.5 ~ 7.5 μM) and compound **88** showed marginal activity against DYRK1A [108].

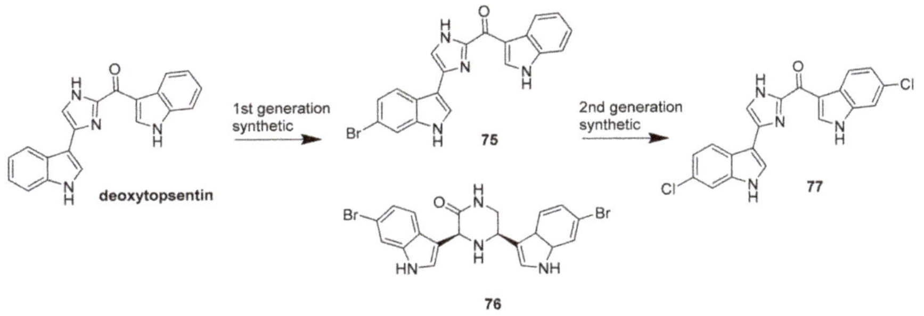

Figure 21. Synthetic analogues (**75–77**) of the marine bisindole alkaloid deoxytopsentin.

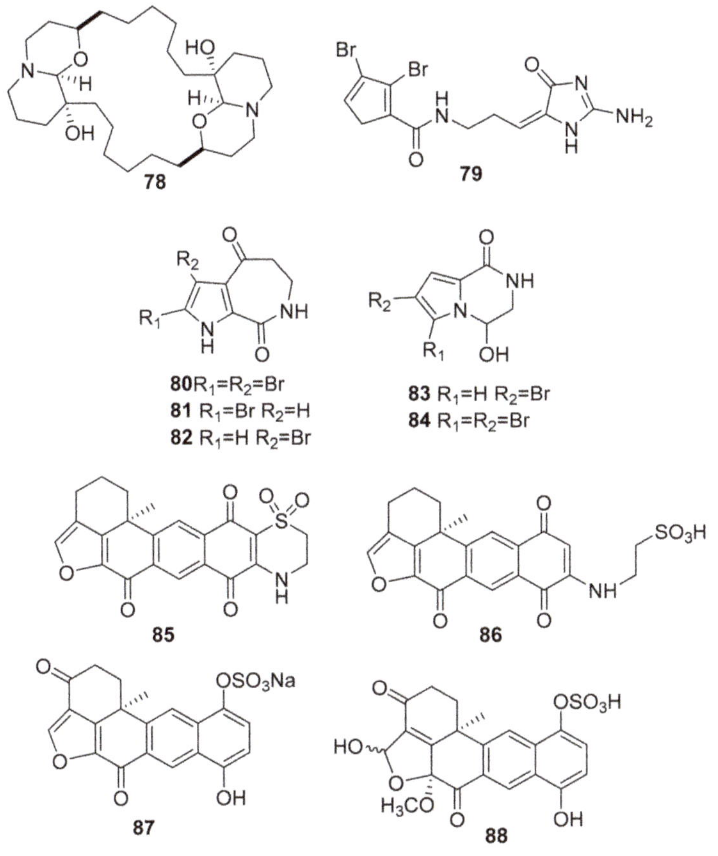

Figure 22. Structure of compounds **78–88**.

In 2015, Esther A. Guzmán et al. [109] reported microsclerodermin A (**89**, Figure 23), from the marine sponge *Amphibleptula cf. madrepora*, and this compound inhibited GSK3β and pancreatic cancer cell viability in vitro (IC_{50} = 2.4 μM). In 2015, Shou-Ping Shih et al. isolated a marine polycyclic quinone called halenaquinone (compound **90**, Figure 23) from the marine sponge *Petrosia* sp. [110]. This molecule is a broad spectrum tyrosine kinase inhibitor, as well as in vitro cytotoxin against multiple cell types (IC_{50} = 0.18 ~ 8.0 μg/mL) [110]. In 2016, Germana Esposito et al. reported chloromethylhalicyclamine B (**91**, Figure 23) after it was isolated from the sponge *Acanthostrongylophora* sp., and suggested that it "can efficiently interact with the ATP-binding site of CK1δ (IC_{50} = 6 μM) in spite of its globular structure, very different from the planar structure of known inhibitors of CK1δ [111]. As the authors also suggested, this molecule can be a lead molecule for the creation of new CK1δ/ε inhibitors. In 2016, Ran Wang et al. discovered stellettin B (**92**, Figure 23) from the marine sponge *Jaspis stellifera*, and it inhibited phosphorylation of PI3K and Akt [112]. Moreover, compound **92** exhibited potent activitivity against human glioblastoma cancer SF295 cells (IC_{50} = 0.01 μM).

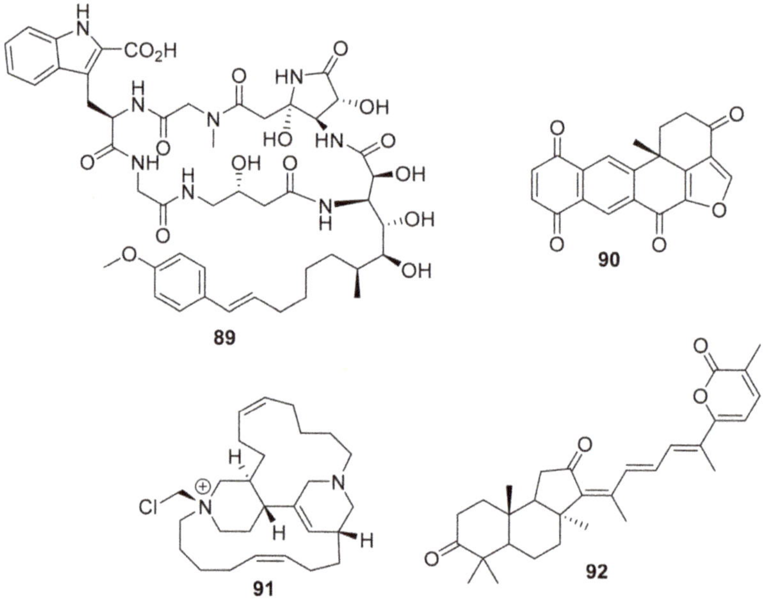

Figure 23. Structure of compounds **89**–**92**.

In 2016, María Roel et al. reported that crambescidins 816, 830, and 800 (compounds **93**–**95**, Figure 24) were isolated from the marine sponge *Smenospongia* sp. [113]. These compounds inhibited tumor cell proliferation by suppressing CDK 2/6 expression and activated the cell CDK inhibitors -2A, -2D and -1A, suggesting some potential as anticancer agents [113]. In 2017, Nadège Loaëc et al. reported polyandrocarpamines A and B (**96** and **97**, Figure 24) as synthetic analogues of the 2-aminoimidazolin-4-one scaffold previously isolated from marine calcareous sponges *Leucetta* and *Clathrina* [114]. Compounds **96** and **97** were found to be nM potency inhibitors of DYRKs (IC_{50} = 0.17 ~ 0.88 μM) and CLKs (IC_{50} = 0.32 ~ 8.6 μM), and the 2-aminoimidazolone scaffold was accordingly proposed as having promise for developing new kinase inhibitors [114].

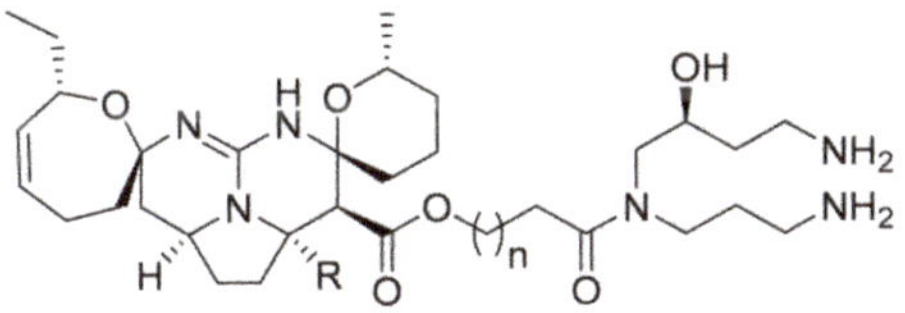

92 n = 14 R = OH
93 n = 15 R = OH
94 n = 14 R = H

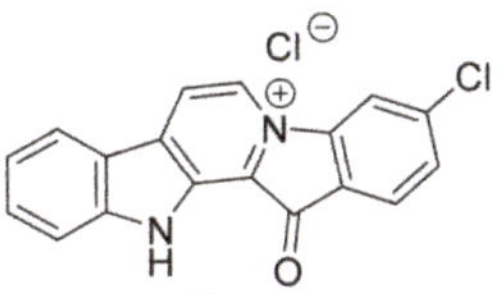

95

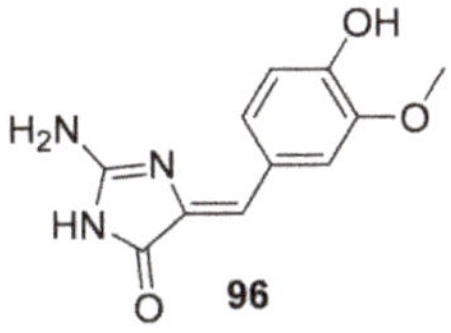

96

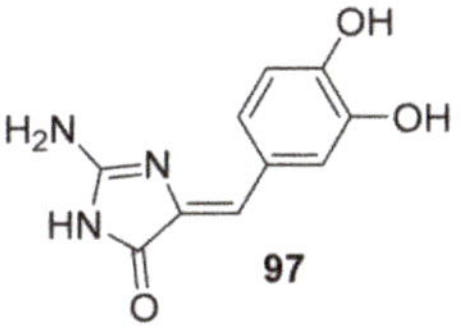

97

Figure 24. Structures of compounds **92–97**.

In 2018, the manzamine alkaloids were isolated from Indo-Pacific marine sponges *Acanthostrongylophora* sp., including manzamine A (**98**), 8-hydroxymanzamine A (**99**), manzamine E (**100**), manzamine F (**101**), 6-deoxymanzamine X (**102**), and the synthetic analogue 6-cyclohexamidomanzamine A (**103**) was also generated [115,116]. These compounds (Figure 25) showed mixed noncompetitive inhibition of MtSK, and the synthetic compound **103** takes advantage of the structural features to pursue applications of anti-inflammatory, antiparasitic, insecticidal, and antibacterial activities with potential for development against malaria and tuberculosis [115,116]. In 2014, Nag S. Kumar et al. reported the discovery of a series of bis-indole alkaloids based on the structure of a marine alkaloid isolated from the marine sponge *Topsentia pachastrelloides*, of which compound **104** (Figure 26) is a promising PK inhibitor [117].

98 $R_1 = R_2 = H$
99 $R_1 = OH$ $R_2 = H$

100 $R_3 = H$
101 $R_3 = OH$

102

103 $R_1 = H$ $R_2 =$

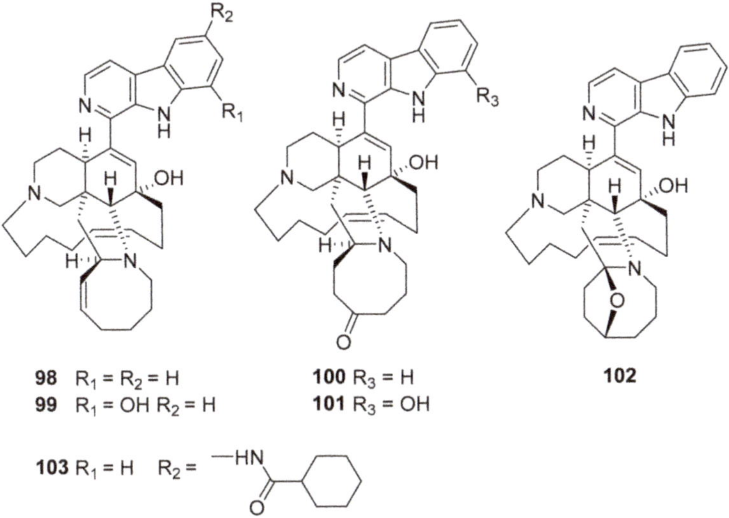

Figure 25. Structures of compounds **98–103**.

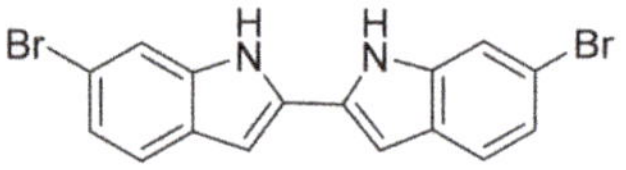

Figure 26. Structure of the bis-indole compound, **104**.

In 2015, Yongseok Kwon et al. synthesized compound **105** (Scheme 4) as a synthetic analogue of pachastrissamine (first reported in 2002 from the marine sponge *Pachastrissa* sp.), and it has in vitro cytotoxicity along with sphingosine kinase inhibitory activity that was deemed superior to the natural lead compound [118,119]. The IC_{50} of compound **105** and pachastrissamine in the inhibition of SphK1 and SphK2 were 7.5 μM and 12.0 μM, 20.1 μM and 41.8 μM, respectively, suggesting that compound **105** could have preferable usage in related pharmaceutical applications [118,119]. The synthetic route of the pachastrissamine carbocyclic analogue **105** is depicted in Scheme 4. The compound (+)-liphagal (**106**, Scheme 5) was first isolated in 2006 from a marine sponge by Frederick and Andersen et al. [120] and in 2015 the chemical synthesis was completed by Markwell-Heys et al. [121]. This compound inhibited PIK-α (IC_{50} = 100 nM), and was also found to be cytotoxic to several tumor cell lines with IC_{50} ≈ 1 μM. It has been suggested that **106** has potential application as a new type of kinase inhibitor or even cancer drug [122]. The synthetic route of (+)-liphagal is depicted in Scheme 5, which initiates with (+)-sclareolide (commercially available) and generates the final product in ~10% yield after 13 reaction steps [122].

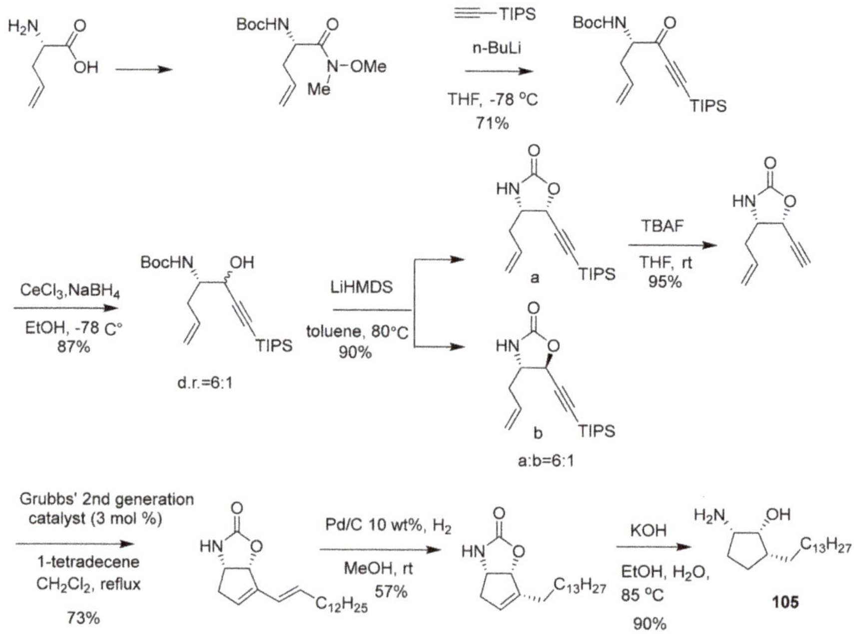

Scheme 4. Synthetic route and structure of pachastrissamine carbocyclic analogue **105**.

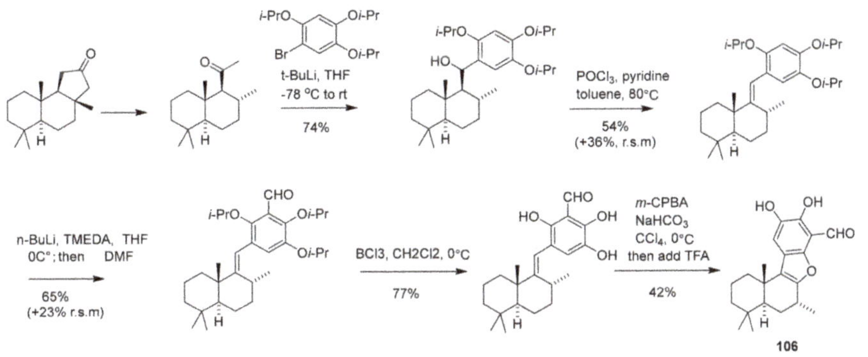

Scheme 5. Synthesis and structure of (+)-liphagal (**106**).

2.2. Preclinical and Clinical Candidates

The US FDA has already approved more than 34 small molecule kinase inhibitors for human use [123]. Many of these have been used for the treatment of cancers, although toxicity is of great concern. A number of marine-derived kinase inhibitors have entered clinic trials, and some are already in later pre-approval stages or have been recently approved (Figures 27 and 28 and Table 7). For example, plitidepsin (aplidin) is a marine natural product that was developed by the Spanish company PharmaMar as an inhibitor of EGPR, Src, JNK and p38 MAPK and, although there were dose-limiting toxicities in some cases, it was not fully withdrawn from clinical trials during phase II [124–126]. Later, it was approved for clinical use in Australia in 2018 for the treatment of relapsed/refractory multiple myeloma patients. Midostaurin is among the important drugs for aggressive systemic mastocytosis (ASM) treatment as a multi-target protein kinase inhibitor, and it was FDA-approved in 2017 [124,127–129]. Protein kinase inhibitors from the marine origin that have entered clinical trials include three kinase inhibitors (lestaurtinib [130], enzastaurin [131], CEP-1347 [14,124,127]) in phase III, three kinase inhibitors (kahalalide F [132,133], 7-hydroxystaurosporine [134,135], staurosporine [136]) in phase II and two kinase inhibitors (CEP-2563 [14,124,127], isokahalalide F [14,124,127]) in phase I clinical studies. Somewhat related, bryostatin 1 entered phase II clinical trials for testing in the treatment of Alzheimer's Disease as a PKC activator, after it was also experimentally used in phase I trials for various forms of cancer [137–139]. It will be interesting to see the future developments of this marine natural product drug candidate and others.

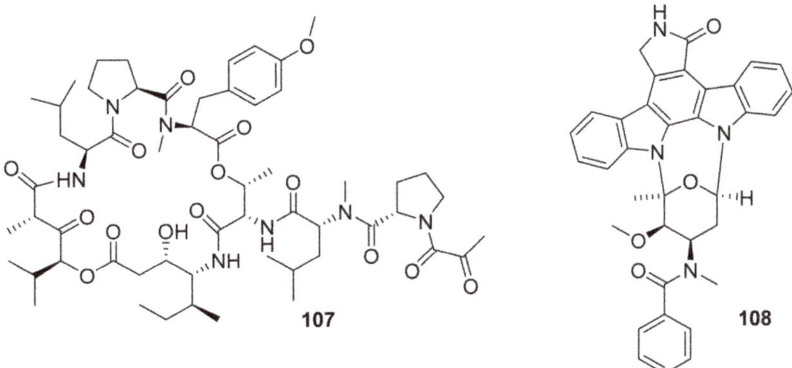

Figure 27. Structures of marine-derived kinase inhibitors in clinical use (**107–108**).

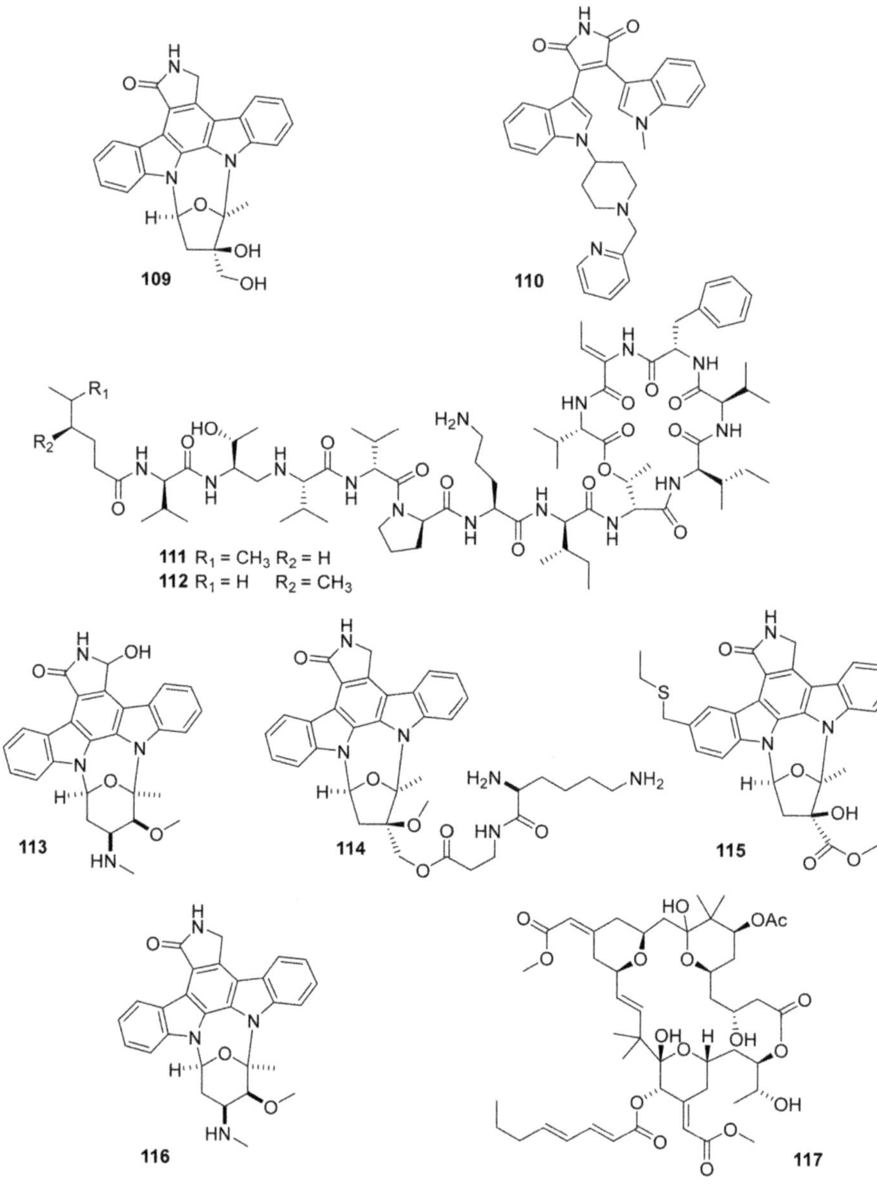

111 $R_1 = CH_3$ $R_2 = H$
112 $R_1 = H$ $R_2 = CH_3$

Figure 28. Structures of marine-derived kinase inhibitors, **109–117**, currently in clinical trials.

Table 7. Preclinical and clinical status of some kinase inhibitors, and bryostatin 1, with marine-derived natural product origins [a,b].

Compound Name (Trademark)	Target	Marine Source	Chemical Class	Company or Institution	Therapeutic Area	Status	References
plitidepsin (aplidin, **107**)	VEGFR1	ascidian	depsipeptide	PharmaMar	multiple myeloma	FDA-approved (2018)	[124–127]
midostaurin (**108**)	PKC-α, KIT, FLT3, VEGFR2, PDGFR-β	ascidian / bacteria	indolocarbazole	Novartis	acute myeloid leukemia,	FDA-approved (2017)	[124,127,128]
lestaurtinib (CEP-701, **109**)	Flt-3, JAK-2	ascidian / bacteria	indolocarbazole	Kyowa kirin	acute lymphoblast leukemia	Phase III	[127,130]
enzastaurin (LY317615, **110**)	PKCβ, Class I PI3K	ascidian / bacteria	indolocarbazole	Eli Lilly	diffuse large B cell lymphoma	Phase III	[127,131]
kahalalide F (PM-92102, **111**)	EGFR	mollusk and algae	tridecapeptide	PharmaMar Cephalon	cancar	Phase II (terminated)	[127,132,133]
isokahalalide F (PM02734, elisidepsin, **112**)	ErbB2	mollusk and algae	tridecapeptide	PharmaMar	cancer	Phase I (terminated)	[14,124,127]
7-hydroxystaurosporine (UCN-01, **113**)	Chk2, PDPK1, Chk1 PKC	ascidian / bacteria	indolocarbazole	Kyowa kirin	cancer	Phase II (terminated)	[127,134,135]
CEP-2563 (**114**, prodrug of CEP-751)	Trk-A, Trk-B, Trk-C	ascidian / bacteria	indolocarbazole	Cephalon	cancer	Phase I (terminated)	[14,124,127]
CEP-1347 (KT7515; **115**)	MAP3K11	ascidian	indolocarbazole	Teva	Parkinson's	Phase III (terminated)	[14,124,127]
staurosporine (AM-2282, **116**)	PKC, JAK2, CamKIII	ascidian / bacteria	indolocarbazole	Kyowa Hakko Kirin (originator)	multiple myeloma	Phase II (terminated)	[127,136]
bryostatin 1 [b] (**117**)	PKC	bryozoan	macrocytic lactone	Neurotrope Bioscience	Alzheimer's	Phase II (terminated)	[127,137–139]

[a] Staurosporine and several derivatives are included here. These compounds have been derived from terrestrial soil bacteria, ascidians and marine bacteria. [b] bryostatin 1 is a Protein Kinase C (PKC) activator.

3. Conclusions

A wide variety of kinase inhibitors have been recently published from marine-derived sources, including bacteria and cyanobacteria, fungi, animals, algae, soft corals and sponges. More than 100 kinase inhibitors of marine origin have been systematically documented for facile examination and comparison. An evaluation of the number of compounds reported during the period of this review, separated by the class of source organism, shows that there is a vast opportunity area for a kinase inhibitor research on the natural product chemistry of marine soft corals, especially (Figure 29). However, it should also be considered that from the overwhelming 58 kinase inhibitors reported from marine animals, soft corals, and sponges during this period, many are likely produced by symbiotic microorganisms. It is equally important to realize, however, that those symbionts may be difficult or impossible to cultivate in laboratory conditions. It is thus of increasing importance that the development of selective kinase inhibitors from marine natural product drug discovery leads has utilized active fragment (pharmacophore) based medicinal chemistry to overcome supply limitations of natural product drug leads, and further optimize the drug-like properties of these molecules. In this way, each newly reported compound may provide marine-derived scaffolds and bioactivity information for drug screening, and many have been found to have significant biological properties that could be used in the treatment of different conditions such as cancers, inflammation, and neurodegenerative diseases, for example.

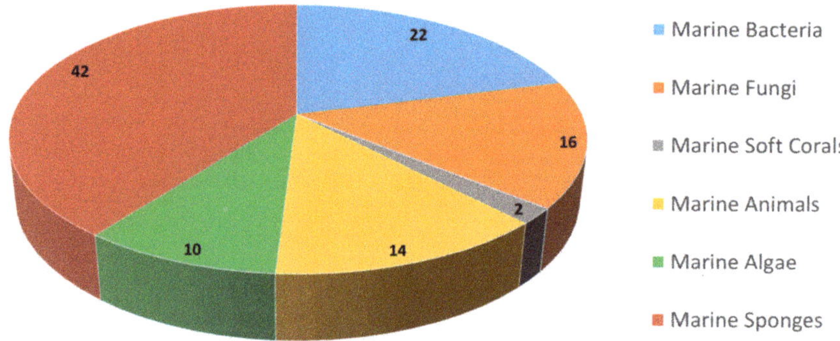

Figure 29. A breakdown by class of the reported producing organism for the marine derived natural product kinase inhibitors covered in this review period, from January, 2014 to February, 2019.

The noted structural differences and diversity of the marine natural products and their derivatives are important for chemical properties, conferring pharmacological effects, and of late-stage interest, intellectual property management. Some recent marine drug approvals include plitidepsin (aplidin), an ascidian natural product, and midostaurin, which is a synthetic derivative of the marine and terrestrial microbial molecule, staurosporine. There is also a pipeline of clinical trial candidates and preclinical development projects behind these drugs.

In recent years, new techniques including deep sea sampling, advanced methods in chemical synthesis, more efficient target-based isolation [140], along with computational database mining strategies and high-throughput screenings have been widely helpful for drug candidate discovery and design. These all improve the efficiency and success rate of the drug development process. Since kinases are directly implicated in so many aspects of disease development and progression in growth, successful targeting of these proteins has resulted in a number of novel drugs for cancer and neurodegenerative diseases and is expected to continue to do so. Nevertheless, while most of the approved kinase inhibitors are ATP binding site competitors, there is a dire need to overcome the toxicity associated with this liability. Hence, marine-derived natural product kinase inhibitors could serve as lead compounds for development by medicinal chemistry. These diverse compounds greatly improve the chemical space covered during screening for new kinase inhibitors, with the goal of eventually providing new clinical candidates for unmet medical needs.

Funding: This research and the APC was funded by the National Key Research and Development Program of China (2018YFC0310900), the National Natural Science Foundation of China (81850410553, 41776168), the National 111 Project of China (D16013), the Li Dak Sum Yip Yio Chin Kenneth Li Marine Biopharmaceutical Development Fund, and the K.C. Wong Magna Fund in Ningbo University, the Natural Science Foundation of Ningbo City (2018A610410) and Foundation of Ningbo University for Grant (XYL18004).

Conflicts of Interest: The authors declare no conflict of interest.

List of Abbreviations

AKt	Serine/ threonine kinase
BTK	Brution tyrosine kinase
BrK	Breast tumor kinase
CLK1	Cdc2-like kinase 1
CLKs	Cdc2-like kinases
CDK1	Cyclin-dependent kinase 1
CDK2	Cyclin-dependent kinase 2
CDK4	Cyclin-dependent kinase 4
CDK5	Cyclin-dependent kinase 5
CDK6	Cyclin-dependent kinase 6
CDK9	Cyclin-dependent kinase 9
CK1	Casein kinase 1
CK5	Casein kinase 5
CAMK1a	Calcium/calmodulin-dependent protein kinase 1a
CDKN1C	Cyclin-dependent kinase inhibitor 1C
CDKN2B	Cyclin-dependent kinase inhibitor 2B
CHEK2	Checkpoint kinase 2
Class I PI3K	PI3-kinase class I
Chk	Csk homologous kinase
CaMKII	Calmodulin-dependent protein kinase II
DYRK1A	Dual-specificity tyrosine phosphorylation regulated kinase-1A
DYRKS	Dual-specificity tyrosine phosphorylation regulated kinases
ERK	Extracellular signal-regulated kinase
ERK1/2	Extracellular signal-regulated kinase1/2
EGFR	Epidermal growth factor receptor
FAK	Focal adhesion kinase
FGER	Fibroblast Growth Factor Receptor
GSK-3	Glycogen synthase kinase 3
GSK-3β	Glycogen synthase kinase 3 beta
IKK	IkappaB kinase
IGF1-R	Insulin-like growth factor 1 receptor
JNK	C-Jun NH-terminal kinase
LIMK	LIM kinase
MtSK	Shikimate kinase from mycobacterium tuberculosis
MRSA PK	MRSA pyruvate kinase
MAPK	Mitogen-activated protein kinases
MAP3K11	Mitogen-activated protein kinase kinase kinase11
MIC	Minimum inhibitory concentration
NF-κB	Nuclear factor – kappa B
PKC-α	Protein kinase C alpha
PKC-β	Protein kinase C beta
PKnG	Protein kinase G

PKB	Protein kinase B
p-AKt	Phosphorylated AKt
p-ERK1/2	Phosphorylated extracellular signal-regulated kinase
PAK1	p21-activated kinase 1
PI3K	Phosphatidylinositol 3-kinase
PDGFR-α	Platelet-derived growth factor receptor alpha
PDGFR-β	Platelet-derived growth factor receptor beta
PDPK1	Phosphatidylinositide-dependent protein kinase 1
ROCK2	Rho-associated protein kinase
RTKs	Receptor tyrosine kinase
RIPK2	Receptor-interacting serine/threonine kinase 2
STK	Serine/threonine kinase family
SIK2	Salt-inducible kinase2
SphKs	Sphingosine kinases
STAT3	Signal transducer and activator of transcription 3
VEGFA	Vascular endothelial growth factor A
VEGFR2	Vascular endothelial growth factor receptor 2
JAK2	Tyrosine-protein kinase JAK2

References

1. Shang, J.; Hu, B.; Wang, J.; Zhu, F.; Kang, Y.; Li, D.; Sun, H.; Kong, D.X.; Hou, T. A cheminformatic insight into the differences between terrestrial and marine originated natural products. *J. Chem. Inf. Model.* **2018**, *58*, 1182–1193. [CrossRef] [PubMed]
2. Pye, C.R.; Bertin, M.J.; Lokey, R.S.; Gerwick, W.H.; Linington, R.G. Retrospective analysis of natural products provides insights for future discovery trends. *Proc. Natl. Acad. Sci. USA* **2017**, *114*, 5601–5606. [CrossRef] [PubMed]
3. Newman, D.J.; Cragg, G.M. Natural products as sources of new drugs over the 30 years from 1981 to 2010. *J. Nat. Prod.* **2012**, *75*, 311–335. [CrossRef] [PubMed]
4. Hertweck, C. Natural products as source of therapeutics against parasitic diseases. *Angew. Chem. Int. Ed.* **2015**, *54*, 14622–14624. [CrossRef] [PubMed]
5. Molinski, T.F.; Dalisay, D.S.; Lievens, S.L.; Saludes, J.P. Drug development from marine natural products. *Nat. Rev. Drug Discov.* **2009**, *8*, 69. [CrossRef] [PubMed]
6. Newman, D.J.; Giddings, L.-A. Natural products as leads to antitumor drugs. *Phytochem. Rev.* **2014**, *13*, 123–137. [CrossRef]
7. Watve, M.G.; Tickoo, R.; Jog, M.M.; Bhole, B.D. How many antibiotics are produced by the genus Streptomyces? *Phytochem. Rev.* **2001**, *176*, 386–390. [CrossRef]
8. El-Elimat, T.; Zhang, X.; Jarjoura, D.; Moy, F.J.; Orjala, J.; Kinghorn, A.D.; Pearce, C.J.; Oberlies, N.H. Chemical diversity of metabolites from fungi, cyanobacteria, and plants relative to FDA-approved anticancer agents. *ACS Med. Chem. Lett.* **2012**, *3*, 645–649. [CrossRef]
9. Feher, M.; Schmidt, J.M. Property distributions: Differences between drugs, natural products, and molecules from combinatorial chemistry. *Chem. Inf. Comput.* **2003**, *43*, 218–227. [CrossRef]
10. Blunt, J.W.; Copp, B.R.; Hu, W.P.; Munro, M.H.; Northcote, P.T.; Prinsep, M.R. Marine natural products. *Nat. Prod. Rep.* **2008**, *25*, 35–94. [CrossRef]
11. Rocha, L.A.; Pinheiro, H.T.; Shepherd, B.; Papastamatiou, Y.P.; Luiz, O.J.; Pyle, R.L.; Bongaerts, P. Mesophotic coral ecosystems are threatened and ecologically distinct from shallow water reefs. *Science* **2018**, *361*, 281–284. [CrossRef] [PubMed]
12. Pereira, F.; Airesdesousa, J. Computational Methodologies in the Exploration of Marine Natural Product Leads. *Mar. Drugs* **2018**, *16*, 236. [CrossRef] [PubMed]
13. Kuttruff, C.A.; Eastgate, M.D.; Baran, P.S. Natural product synthesis in the age of scalability. *Nat. Prod. Rep.* **2014**, *31*, 419–432. [CrossRef] [PubMed]
14. Bharate, S.B.; Sawant, S.D.; Singh, P.P.; Vishwakarma, R.A. Kinase inhibitors of marine origin. *Chem. Rev.* **2013**, *113*, 6761–6815. [CrossRef] [PubMed]

15. Grant, S.K. Therapeutic protein kinase inhibitors. *Cell. Mol. Life Sci.* **2009**, *66*, 1163–1177. [CrossRef] [PubMed]
16. Patterson, H.; Nibbs, R.; Mcinnes, I.; Siebert, S. Protein kinase inhibitors in the treatment of inflammatory and autoimmune diseases. *Clin. Exp. Immunol.* **2014**, *176*, 1–10. [CrossRef]
17. Plowman, G.D.; Hunter, T.; Sudarsanam, S. Evolution of protein kinase signaling from yeast to man. *Trends Biochem.Sci.* **2002**, *27*, 514–520.
18. Ubersax, J.; Ferrell, J. Mechanisms of specificity in protein phosphorylation. *Nat. Rev. Mol. Cell Biol.* **2007**, *8*, 530–541. [CrossRef]
19. Official Website of U.S. Food and Drug Administration. Available online: https://www.fda.gov/drugs/development-approval-process-drugs/new-drugs-fda-cders-new-molecular-entities-and-new-therapeutic-biological-products (accessed on 10 April 2019).
20. Ferguson, F.M.; Gray, N.S. Kinase inhibitors: The road ahead. *Nat. Rev. Drug Discov.* **2018**, *17*, 353. [CrossRef]
21. Rowinsky, E.K. The erbB family: Targets for therapeutic development against cancer and therapeutic strategies using monoclonal antibodies and tyrosine kinase inhibitors. *Annu. Rev. Med.* **2004**, *55*, 433. [CrossRef]
22. Yao, H.P.; Zhou, Y.Q.; Ma, Q.; Guin, S.; Padhye, S.S.; Zhang, R.W.; Wang, M.H. The monoclonal antibody Zt/f2 targeting RON receptor tyrosine kinase as potential therapeutics against tumor growth-mediated by colon cancer cells. *Mol. Cancer* **2011**, *10*, 82. [CrossRef] [PubMed]
23. Hassan, A.Q.; Sharma, S.V.; Markus, W. Allosteric inhibition of BCR-ABL. *Cell Cycle* **2010**, *9*, 3734–3738. [CrossRef] [PubMed]
24. Hardy, J.A.; Wells, J.A. Searching for new allosteric sites in enzymes. *Curr. Opin. Struct. Biol.* **2004**, *14*, 706–715. [CrossRef] [PubMed]
25. Lamba, V.; Ghosh, I. New directions in targeting protein kinases: Focusing upon true allosteric and bivalent inhibitors. *Curr. Pharm. Des.* **2012**, *18*, 706–715. [CrossRef] [PubMed]
26. Schenone, S.; Brullo, C.; Musumeci, F.; Radi, M.; Botta, M. ATP-competitive inhibitors of mTOR: An update. *Curr. Med. Chem.* **2011**, *18*, 2995–3014. [CrossRef] [PubMed]
27. Bonnet, P.; Mucs, D.; Bryce, R.A. Targeting the inactive conformation of protein kinases: Computational screening based on ligand conformation. *MedChemComm* **2012**, *3*, 434–440. [CrossRef]
28. Dar, A.C.; Shokat, K.M. The Evolution of Protein Kinase Inhibitors from Antagonists to Agonists of Cellular Signaling. *Annu. Rev. Biochem.* **2011**, *80*, 769–795. [CrossRef]
29. Liu, Y.; Gray, N.S. Rational design of inhibitors that bind to inactive kinase conformations. *Nat. Chem. Biol.* **2006**, *2*, 358–364. [CrossRef]
30. Simard, J.R.; Pawar, V.; Aust, B.; Wolf, A.; Rabiller, M.; Wulfert, S.; Robubi, A.; Klüter, S.; Ottmann, C.; Rauh, D. High-Throughput Screening To Identify Inhibitors Which Stabilize Inactive Kinase Conformations in p38α. *J. Am. Chem. Soc.* **2009**, *51*, 18478–18488. [CrossRef]
31. Oleg, F.; Brian, M.; Vanda, P.; Peter, R.; Susanne, M.; Bullock, A.N.; Juerg, S.; Michael, S.M.; Stefan, K. A systematic interaction map of validated kinase inhibitors with Ser/Thr kinases. *Proc. Natl. Acad. Sci. USA* **2007**, *104*, 20523–20528.
32. Website of Cell Signaling Technology, Inc. Available online: https://media.cellsignal.com/www/images/science/kinases/kinome.jpg (accessed on 1 May 2019).
33. Dey, G.; Bharti, R.; Dhanarajan, G.; Das, S.; Dey, K.K.; Kumar, B.N.P.; Sen, R.; Mandal, M. Marine lipopeptide Iturin A inhibits Akt mediated GSK3β and FoxO3a signaling and triggers apoptosis in breast cancer. *Sci. Rep.* **2015**, *5*, 10316. [CrossRef] [PubMed]
34. Zhou, B.; Qin, L.L.; Ding, W.J.; Ma, Z.J. Cytotoxic indolocarbazoles alkaloids from the *streptomyces* sp. A65. *Tetrahedron* **2018**, *74*, 726–730. [CrossRef]
35. Qin, L.L.; Zhou, B.; Ding, W.; Ma, Z. Bioactive metabolites from marine-derived *Streptomyces* sp. A68 and its Rifampicin resistant mutant strain R-M1. *Phytochem. Lett.* **2018**, *23*, 46–51. [CrossRef]
36. Jiang, Y.J.; Li, J.Q.; Zhang, H.J.; Ding, W.J.; Ma, Z.J. Cyclizidine-Type Alkaloids from *Streptomyces* sp. HNA39. *J. Nat. Prod.* **2018**, *81*, 394–399. [CrossRef] [PubMed]
37. Freer, A.A.; Gardner, D.; Greatbanks, D.; Poyser, J.P.; Sim, G.A. Structure of cyclizidine (antibiotic M146791): X-ray crystal structure of an indolizidinediol metabolite bearing a unique cyclopropyl side-chain. *J. Chem. Soc. Chem. Commun.* **1982**, *20*, 1160–1162. [CrossRef]

38. Gomi, S.; Ikeda, D.; Nakamura, H.; Naganawa, H.; Yamashita, F.; Hotta, K.; Kondo, S.; Okami, Y.; Umezawa, H.; Iitaka, Y. Isolation and structure of a new antibiotic, indolizomycin, produced by a strain SK2-52 obtained by interspecies fusion treatment. *J. Antibiot.* **1984**, *37*, 1491–1494. [CrossRef] [PubMed]
39. Miho, I.; Takahiro, H.; Motoki, T.; Kazuo, S.Y. A new cyclizidine analog-JBIR-102-from *Saccharopolyspora* sp. RL78 isolated from mangrove soil. *J. Antibiot.* **2012**, *65*, 41.
40. Wang, J.N.; Zhang, H.J.; Li, J.Q.; Ding, W.J.; Ma, Z.J. Bioactive Indolocarbazoles from the Marine-Derived *Streptomyces* sp. DT-A61. *J. Nat. Prod.* **2018**, *81*, 949–956. [CrossRef] [PubMed]
41. Cheng, X.; Zhou, B.; Liu, H.; Huo, C.; Ding, W. One new indolocarbazole alkaloid from the *Streptomyces* sp. A22. *Nat. Prod. Res.* **2018**, *32*, 2583–2588. [CrossRef]
42. Zhou, B.; Hu, Z.J.; Zhang, H.J.; Li, J.Q.; Ding, W.J.; Ma, Z.J. Bioactive staurosporine derivatives from the *Streptomyces* sp. NB-A13. *Bioorg. Chem.* **2018**, *82*, 33–40. [CrossRef]
43. Jiang, Y.-J.; Zhang, D.-S.; Zhang, H.-J.; Li, J.-Q.; Ding, W.-J.; Xu, C.-D.; Ma, A.Z.-J. Medermycin-Type Naphthoquinones from the Marine-Derived *Streptomyces* sp. XMA39. *J. Nat. Prod.* **2018**, *9*, 2120–2124. [CrossRef] [PubMed]
44. Chen, D.; Ma, S.; He, L.; Yuan, P.; She, Z.; Lu, Y. Sclerotiorin inhibits protein kinase G from *Mycobacterium tuberculosis* and impairs mycobacterial growth in macrophages. *Curr. Top. Microbiol. Immunol.* **2017**, *103*, 37–43. [CrossRef] [PubMed]
45. Dong-Cheol, K.; Hee-Suk, L.; Wonmin, K.; Dong-Sung, L.; Jae Hak, S.; Joung Han, Y.; Youn-Chul, K.; Hyuncheol, O. Anti-inflammatory effect of methylpenicinoline from a marine isolate of *Penicillium* sp. (SF-5995): Inhibition of NF-κB and MAPK pathways in lipopolysaccharide-induced RAW264.7 macrophages and BV2 microglia. *Molecules* **2014**, *19*, 18073–18089.
46. Li, W.; Li, M.; Su, X.; Qin, L.; Miao, M.; Yu, C.; Shen, Y.; Luo, Q.; Chen, Q. Mycoepoxydiene induces apoptosis and inhibits TPA-induced invasion in human cholangiocarcinoma cells via blocking NF-κB pathway. *Biochimie* **2014**, *101*, 183–191. [CrossRef] [PubMed]
47. Sun, K.; Li, Y.; Guo, L.; Wang, Y.; Liu, P.; Zhu, W. Indole diterpenoids and isocoumarin from the fungus, *Aspergillus flavus*, isolated from the prawn, Penaeus vannamei. *Mar. Drugs* **2014**, *12*, 3970–3981. [CrossRef] [PubMed]
48. Chen, W.L.; Turlova, E.; Sun, C.L.F.; Kim, J.S.; Huang, S.; Zhong, X.; Guan, Y.Y.; Wang, G.L.; Rutka, J.T.; Feng, Z.P. Xyloketal B Suppresses Glioblastoma Cell Proliferation and Migration in Vitro through Inhibiting TRPM7-Regulated PI3K/Akt and MEK/ERK Signaling Pathways. *Mar. Drugs* **2015**, *13*, 2505–2525. [CrossRef] [PubMed]
49. Kim, D.C.; Quang, T.H.; Ngan, N.T.; Yoon, C.S.; Sohn, J.H.; Yim, J.H.; Feng, Y.; Che, Y.; Kim, Y.C.; Oh, H. Dihydroisocoumarin Derivatives from Marine-Derived Fungal Isolates and Their Anti-inflammatory Effects in Lipopolysaccharide-Induced BV2 Microglia. *J. Nat. Prod.* **2015**, *78*, 2948–2955. [CrossRef] [PubMed]
50. Kim, J.W.; Ko, S.K.; Kim, H.M.; Kim, G.H.; Son, S.; Kim, G.S.; Hwang, G.J.; Jeon, E.S.; Shin, K.S.; Ryoo, I.J. Stachybotrysin, an Osteoclast Differentiation Inhibitor from the Marine-Derived Fungus *Stachybotrys* sp. KCB13F013. *J. Nat. Prod.* **2016**, *79*, 2703–2708. [CrossRef] [PubMed]
51. Wiese, J.; Imhoff, J.F.; Gulder, T.A.; Labes, A.; Schmaljohann, R. Marine Fungi as Producers of Benzocoumarins, a New Class of Inhibitors of Glycogen-Synthase-Kinase 3β. *Mar. Drugs* **2016**, *14*, 200. [CrossRef]
52. Ko, W.; Sohn, J.H.; Jang, J.H.; Ahn, J.S.; Kang, D.G.; Lee, H.S.; Kim, J.S.; Kim, Y.C.; Oh, H. Inhibitory effects of alternaramide on inflammatory mediator expression through TLR4-MyD88-mediated inhibition of NF-κB and MAPK pathway signaling in lipopolysaccharide-stimulated RAW264.7 and BV2 cells. *Chem. Biol. Interact.* **2016**, *244*, 16–26. [CrossRef] [PubMed]
53. Cho, K.H.; Kim, D.C.; Yoon, C.S.; Ko, W.M.; Lee, S.J.; Sohn, J.H.; Jang, J.H.; Ahn, J.S.; Kim, Y.C.; Oh, H. Anti-neuroinflammatory effects of citreohybridonol involving TLR4-MyD88-mediated inhibition of NF-κB and MAPK signaling pathways in lipopolysaccharide-stimulated BV2 cells. *Neurochem. Int.* **2016**, *95*, 55–62. [CrossRef]
54. García-Caballero, M.; Blacher, S.; Paupert, J.; Quesada, A.R.; Medina, M.A.; Noël, A. Novel application assigned to toluquinol: Inhibition of lymphangiogenesis by interfering with VEGF-C/VEGFR-3 signaling pathway. *Br. J. Pharmacol.* **2016**, *173*, 1966–1987. [CrossRef] [PubMed]
55. Wu, B.; Wiese, J.; Schmaljohann, R.; Imhoff, J. Biscogniauxone, a New Isopyrrolonaphthoquinone Compound from the Fungus *Biscogniauxia mediterranea* Isolated from Deep-Sea Sediments. *Mar. Drugs* **2016**, *14*, 204. [CrossRef] [PubMed]

56. Ngan, N.T.T.; Quang, T.H.; Kim, K.W.; Kim, H.J.; Sohn, J.H.; Kang, D.G.; Lee, H.S.; Kim, Y.C.; Oh, H. Anti-inflammatory effects of secondary metabolites isolated from the marine-derived fungal strain *Penicillium* sp. SF-5629. *Arch. Pharm. Res.* **2017**, *40*, 328–337. [CrossRef] [PubMed]
57. Skropeta, D.; Wei, L. Recent advances in deep-sea natural products. *Nat. Prod. Rep.* **2014**, *31*, 999–1025. [CrossRef] [PubMed]
58. Navarri, M.; Jégou, C.; Bondon, A.; Pottier, S.; Bach, S.; Baratte, B.; Ruchaud, S.; Barbier, G.; Burgaud, G.; Fleury, Y. Bioactive Metabolites from the Deep Subseafloor Fungus *Oidiodendron griseum* UBOCC-A-114129. *Mar. Drugs* **2017**, *15*, 111. [CrossRef] [PubMed]
59. Li, P.C.; Sheu, M.J.; Ma, W.F.; Pan, C.H.; Sheu, J.H.; Wu, C.H. Anti-Restenotic Roles of Dihydroaustrasulfone Alcohol Involved in Inhibiting PDGF-BB-Stimulated Proliferation and Migration of Vascular Smooth Muscle Cells. *Mar. Drugs* **2015**, *13*, 3046–3060. [CrossRef]
60. Mohyeldin, M.M.; Akl, M.R.; Siddique, A.B.; Hassan, H.M.; El Sayed, K.A. The marine-derived pachycladin diterpenoids as novel inhibitors of wild-type and mutant EGFR. *Biochem. Pharmacol.* **2017**, *126*, 51–68. [CrossRef]
61. Hassan, H.M.; Khanfar, M.A.; Elnagar, A.Y.; Mohammed, R.; Shaala, L.A.; Youssef, D.T.; Hifnawy, M.S.; El Sayed, K.A. Pachycladins A-E, prostate cancer invasion and migration inhibitory Eunicellin-based diterpenoids from the red sea soft coral *Cladiella pachyclados*. *J. Nat. Prod.* **2010**, *73*, 848. [CrossRef]
62. Mohamed, G.A.; Ibrahim, S.R.M.; Badr, J.M.; Youssef, D.T.A. Didemnaketals D and E, bioactive terpenoids from a Red Sea ascidian *Didemnum* species. *Tetrahedron* **2014**, *45*, 35–40. [CrossRef]
63. Youssef, D.T.A.; Mohamed, G.A.; Shaala, L.A.; Badr, J.M.; Bamanie, F.H.; Ibrahim, S.R.M. New purine alkaloids from the Red Sea marine tunicate Symplegma rubra. *Phytochem. Lett.* **2015**, *13*, 212–217. [CrossRef]
64. Adrian, T.E.; Collin, P. The Anti-Cancer Effects of Frondoside A. *Mar. Drugs* **2018**, *16*, 64. [CrossRef] [PubMed]
65. Bcq, N.; Yoshimura, K.; Kumazawa, S.; Tawata, S.; Maruta, H. Frondoside A from sea cucumber and nymphaeols from Okinawa propolis: Natural anti-cancer agents that selectively inhibit PAK1 in vitro. *Drug Discov. Ther.* **2017**, *11*, 110–114.
66. Wätjen, W.; Ebada, S.S.; Bergermann, A.; Chovolou, Y.; Totzke, F.; Kubbutat, M.H.G.; Lin, W.; Proksch, P. Cytotoxic effects of the anthraquinone derivatives 1′-deoxyrhodoptilometrin and (S)-(−)-rhodoptilometrin isolated from the marine echinoderm *Comanthus* sp. *Arch. Toxicol.* **2016**, *91*, 1485–1495. [CrossRef] [PubMed]
67. Wright, A.D.; Nielson, J.L.; Tapiolas, D.M.; Motti, C.A.; Ovenden, S.P.B.; Kearns, P.S.; Liptrot, C.H. Detailed NMR, Including 1,1-ADEQUATE, and Anticancer Studies of Compounds from the Echinoderm *Colobometra perspinosa*. *Mar. Drugs* **2009**, *7*, 565–575. [CrossRef] [PubMed]
68. Wang, S.; Wang, L.J.; Jiang, B.; Wu, N.; Li, X.; Liu, S.; Luo, J.; Shi, D. Anti-Angiogenic Properties of BDDPM, a Bromophenol from Marine Red Alga *Rhodomela confervoides*, with Multi Receptor Tyrosine Kinase Inhibition Effects. *Int. J. Mol. Sci.* **2015**, *16*, 13548–13560. [CrossRef] [PubMed]
69. Wu, N.; Luo, J.; Jiang, B.; Wang, L.; Wang, S.; Wang, C.; Fu, C.; Li, J.; Shi, D. Marine bromophenol bis (2,3-dibromo-4,5-dihydroxy-phenyl)-methane inhibits the proliferation, migration, and invasion of hepatocellular carcinoma cells via modulating β1-Integrin/FAK Signaling integrin/FAK signaling. *Mar. Drugs* **2015**, *13*, 1010–1025. [CrossRef] [PubMed]
70. Eom, S.H.; Lee, E.H.; Park, K.; Kwon, J.Y.; Kim, P.H.; Jung, W.K.; Kim, Y.M. Eckol from *Eisenia bicyclis* Inhibits Inflammation Through the Akt/NF-kB Signaling in Propionibacterium acnes-Induced Human Keratinocyte Hacat Cells. *J. Food Biochem.* **2016**, *41*, e12312. [CrossRef]
71. Lin, J.; Yu, J.; Zhao, J.; Zhang, K.; Zheng, J.; Wang, J.; Huang, C.; Zhang, J.; Yan, X.; Gerwick, W.H. Fucoxanthin, a Marine Carotenoid, Attenuates β-Amyloid Oligomer-Induced Neurotoxicity Possibly via Regulating the PI3K/Akt and the ERK Pathways in SH-SY5Y Cells. *Oxidat. Med. Cell. Longev.* **2017**, *2017*. [CrossRef]
72. Satomi, Y. Antitumor and Cancer-preventative Function of Fucoxanthin: A Marine Carotenoid. *Anticancer Res.* **2017**, *37*, 1557–1562. [CrossRef] [PubMed]
73. Zhao, D.; Kwon, S.H.; Chun, Y.S.; Gu, M.Y.; Yang, H.O. Anti-Neuroinflammatory Effects of Fucoxanthin via Inhibition of Akt/NF-κB and MAPKs/AP-1 Pathways and Activation of PKA/CREB Pathway in Lipopolysaccharide-Activated BV-2 Microglial Cells. *Neurochem. Res.* **2017**, *42*, 1–11. [CrossRef] [PubMed]
74. Choi, Y.K.; Ye, B.R.; Kim, E.A.; Kim, J.; Kim, M.S.; Lee, W.W.; Ahn, G.N.; Kang, N.; Jung, W.K.; Heo, S.J. Bis (3-bromo-4,5-dihydroxybenzyl) ether, a novel bromophenol from the marine red alga *Polysiphonia morrowii* that suppresses LPS-induced inflammatory response by inhibiting ROS-mediated ERK signaling pathway in RAW 264.7 macrophages. *Biomed. Pharmacother.* **2018**, *103*, 1170–1177. [CrossRef] [PubMed]

75. Paudel, U.; Lee, Y.H.; Kwon, T.H.; Park, N.H.; Yun, B.S.; Hwang, P.H.; Yi, H.K. Eckols reduce dental pulp inflammation through the ERK1/2 pathway independent of COX-2 inhibition. *Oral Dis.* **2015**, *20*, 827–832. [CrossRef] [PubMed]
76. Cai, W.; Chen, Q.Y.; Dang, L.H.; Luesch, H. Apratoxin S10, a Dual Inhibitor of Angiogenesis and Cancer Cell Growth To Treat Highly Vascularized Tumors. *ACS Med. Chem. Lett.* **2017**, *8*, 1007–1012. [CrossRef] [PubMed]
77. Choi, H.; Mascuch, S.J.; Villa, F.A.; Byrum, T.; Gerwick, W.H. Honaucins A-C, Potent Inhibitors of Inflammation and Bacterial Quorum Sensing: Synthetic Derivatives and Structure-Activity Relationships. *Chem. Biol.* **2012**, *19*, 589–598. [CrossRef] [PubMed]
78. Sapkota, M.; Li, L.; Choi, H.; Gerwick, W.H.; Soh, Y. Bromo-honaucin A inhibits osteoclastogenic differentiation in RAW 264.7 cells via Akt and ERK signaling pathways. *Eur. J. Pharmacol.* **2015**, *769*, 100–109. [CrossRef] [PubMed]
79. Pereira, A.; Cao, Z.; Murray, T.F.; Gerwick, W.H. Hoiamide A, a Sodium Channel Activator of Unusual Architecture from a Consortium of Two Papua New Guinea Cyanobacteria. *Chem. Biol.* **2009**, *16*, 893–906. [CrossRef] [PubMed]
80. Cao, Z.; Xichun, L.; Xiaohan, Z.; Michael, G.; William, G.; Thomas, M. Involvement of JNK and Caspase Activation in Hoiamide A-Induced Neurotoxicity in Neocortical Neurons. *Mar. Drugs* **2015**, *13*, 903–919. [CrossRef] [PubMed]
81. Medina, R.A.; Goeger, D.E.; Hills, P.; Mooberry, S.L.; Huang, N.; Romero, L.I.; Ortega-Barri, E.; Gerwick, W.H.; McPhail, K.L. Coibamide A, a Potent Antiproliferative Cyclic Depsipeptide from the Panamanian Marine Cyanobacterium *Leptolyngbya* sp. *J. Am. Chem. Soc.* **2008**, *130*, 6324–6325. [CrossRef]
82. Serrill, J.D.; Wan, X.; Hau, A.M.; Jang, H.S.; Coleman, D.J.; Indra, A.K.; Alani, A.W.G.; Mcphail, K.L.; Ishmael, J.E. Coibamide A, a natural lariat depsipeptide, inhibits VEGFA/VEGFR2 expression and suppresses tumor growth in glioblastoma xenografts. *Investig. New Drugs* **2016**, *34*, 24–40. [CrossRef]
83. Luesch, H.; Yoshida, W.Y.; Moore, R.E.; Paul, V.J.; Corbett, T.H. Total structure determination of apratoxin A, a potent novel cytotoxin from the marine cyanobacterium *Lyngbya majuscula*. *J. Am. Chem. Soc.* **2001**, *123*, 5418–5423. [CrossRef] [PubMed]
84. Fabien, P.; Pritesh, P.; Xue, X.; Piggott, A.M.; Xiao-Cong, H.; Zeinab, K.; Capon, R.J. Callyspongisines A-D: Bromopyrrole alkaloids from an Australian marine sponge, *Callyspongia* sp. *Org. Biomol. Chem.* **2014**, *12*, 1579–1584.
85. Cimino, G.; Rosa, S.D.; Stefano, S.D.; Mazzarella, L.; Puliti, R.; Sodano, G. Isolation and X-ray crystal structure of a novel bromo-compound from two marine sponges. *Tetrahedron Lett.* **1982**, *23*, 767–768. [CrossRef]
86. Williams, D.H.; Faulkner, D.J. Isomers and Tautomers of Hymenialdisine and Debromohymenialdisine. *Nat. Prod. Lett.* **1996**, *9*, 57–64. [CrossRef]
87. Yu, Z.G.; Bi, K.S.; Guo, Y.W. Hyrtiosins A–E, Five New Scalarane Sesterterpenes from the South China Sea Sponge *Hyrtios erecta*. *Helv. Chim. Acta* **2005**, *88*, 1004–1009. [CrossRef]
88. He, W.F.; Xue, D.Q.; Yao, L.G.; Li, J.Y.; Li, J.; Guo, Y.W. Hainanerectamines A–C, alkaloids from the Hainan sponge *Hyrtios erecta*. *Mar. Drugs* **2014**, *12*, 3982–3993. [CrossRef]
89. Akl, M.R.; Foudah, A.I.; Ebrahim, H.Y.; Meyer, S.A.; El Sayed, K.A. The marine-derived sipholenol A-4-O-3′,4′-dichlorobenzoate inhibits breast cancer growth and motility *in vitro* and *in vivo* through the suppression of Brk and FAK signaling. *Mar. Drugs* **2014**, *12*, 2282–2304. [CrossRef]
90. Lidgren, G.; Bohlin, L.; Bergman, J. Studies of swedish marine organisms VII. A novel biologically active indole alkaloid from the sponge *geodia baretti*. *Tetrahedron Lett.* **1986**, *27*, 3283–3284. [CrossRef]
91. Lind, K.F.; Østerud, B.; Hansen, E.; Jørgensen, T.Ø.; Andersen, J.H. The immunomodulatory effects of barettin and involvement of the kinases CAMK1α and RIPK2. *Immunopharmacol. Immunotoxicol.* **2015**, *5*, 458–464. [CrossRef]
92. Lind, K.F.; Espen, H.; Bjarne, S.; Karl-Erik, E.; Annette, B.; Magnus, E.; Kinga, L.; Andersen, J.H. Antioxidant and anti-inflammatory activities of barettin. *Mar. Drugs* **2013**, *11*, 2655–2666. [CrossRef]
93. Rebecca, H.; Boris, P.; Eugen, J.; Joachim, S.; Dorian, S.; Daniel, C.; Frank, T.; Christoph, S.C.; Christian, P. Optimization of potent DFG-in inhibitors of platelet derived growth factor receptorβ (PDGF-Rβ) guided by water thermodynamics. *J. Med. Chem.* **2014**, *58*, 170–182.

94. Segraves, N.L.; Lopez, S.; Johnson, T.A.; Said, S.A.; Fu, X.; Schmitz, F.J.; Crews, P. Structures and cytotoxicities of fascaplysin and related alkaloids from two marine phyla-*Fascaplysinopsis* sponges and *Didemnum* tunicates. *Tetrahedron Lett.* **2003**, *44*, 3471–3475. [CrossRef]
95. Emmanuel, A.; Thoma s, S.T.; Isabelle, M.; Hermann, E.; Menger, M.D.; Laschke, M.W. The Marine-Derived Kinase Inhibitor Fascaplysin Exerts Anti-Thrombotic Activity. *Mar. Drugs* **2015**, *13*, 6774–6791.
96. Mahale, S.; Bharate, S.B.; Manda, S.; Joshi, P.; Jenkins, P.R.; Vishwakarma, R.A.; Chaudhuri, B. Antitumour potential of BPT: A dual inhibitor of cdk4 and tubulin polymerization. *Cell Death Dis.* **2015**, *6*, 1743. [CrossRef] [PubMed]
97. Hamilton, G. Cytotoxic effects of fascaplysin against small cell lung cancer cell lines. *Mar. Drugs* **2014**, *12*, 1377–1389. [CrossRef] [PubMed]
98. Suresh, K.; Santosh Kumar, G.; Anup Singh, P.; Sudhakar, M.; Ajay, K.; Bharate, S.B.; Vishwakarma, R.A.; Fayaz, M.; Shashi, B. Fascaplysin induces caspase mediated crosstalk between apoptosis and autophagy through the inhibition of PI3K/AKT/mTOR signaling cascade in human leukemia HL-60 cells. *J. Cell. Biochem.* **2015**, *116*, 985–997.
99. Sharma, S.; Guru, S.K.; Manda, S.; Kumar, A.; Mintoo, M.J.; Prasad, V.D.; Sharma, P.R.; Mondhe, D.M.; Bharate, S.B.; Bhushan, S. A marine sponge alkaloid derivative 4-chloro fascaplysin inhibits tumor growth and VEGF mediated angiogenesis by disrupting PI3K/Akt/mTOR signaling cascade. *Chem. Biol. Interact.* **2017**, *275*, 47–60. [CrossRef]
100. Veale, C.G.L.; Roya, Z.; Young, R.M.; Morrison, J.P.; Manoja, P.; Lobb, K.A.; Reiner, N.E.; Andersen, R.J.; Davies-Coleman, M.T. Synthetic analogues of the marine bisindole deoxytopsentin: Potent selective inhibitors of MRSA pyruvate kinase. *J. Nat. Prod.* **2015**, *78*, 355–362. [CrossRef]
101. Akl, M.R.; Ayoub, N.M.; Ebrahim, H.Y.; Mohyeldin, M.M.; Orabi, K.Y.; Foudah, A.I.; El Sayed, K.A. Araguspongine C induces autophagic death in breast cancer cells through suppression of c-Met and HER2 receptor tyrosine kinase signaling. *Mar. Drugs* **2015**, *13*, 288–311. [CrossRef]
102. Ebada, S.S.; Hoang, L.M.; Arlette, L.; De Voogd, N.J.; Emilie, D.; Laurent, M.; Marie-Lise, B.K.; Singab, A.N.B.; Müller, W.E.G.; Peter, P. Dispacamide E and other bioactive bromopyrrole alkaloids from two Indonesian marine sponges of the genus *Stylissa*. *Nat. Prod. Res.* **2015**, *29*, 231–238. [CrossRef]
103. Schmitz, F.J.; Gunasekera, S.P.; Lakshmi, V.; Tillekeratne, L.M. Marine natural products: Pyrrololactams from several sponges. *J. Nat. Prod.* **1985**, *48*, 47–53. [CrossRef] [PubMed]
104. Hassan, W.; Elkhayat, E.S.; Edrada, R.A.; Ebel, R.; Proksch, P. New Bromopyrrole Alkaloids From The Marine Sponges *Axinella Damicornis* And *Stylissa Flabelliformis*. *Nat. Prod. Commun.* **2007**, *2*, 1149–1154. [CrossRef]
105. Li, C.J.; Schmitz, F.J.; Kelly-Borges, M. A new lysine derivative and new 3-bromopyrrole carboxylic acid derivative from two marine sponges. *J. Nat. Prod.* **1998**, *61*, 387–389. [CrossRef] [PubMed]
106. Uemoto, H.; Tsuda, M.; Kobayashi, J.I. Mukanadins A–C, New Bromopyrrole Alkaloids from Marine Sponge *Agelas nakamurai*. *J. Nat. Prod.* **1999**, *62*, 1581–1583. [CrossRef] [PubMed]
107. Cafieri, F.; Fattorusso, E.; Mangoni, A.; Taglialatela-Scafati, O. Longamide and 3,7-dimethylisoguanine, two novel alkaloids from the marine sponge *Agelas longissima*. *Tetrahedron Lett.* **2010**, *36*, 7893–7896. [CrossRef]
108. Fei, H.; Mai, L.H.; Longeon, A.; Copp, B.R.; Loaëc, N.; Bescond, A.; Meijer, L.; Bourguet-Kondracki, M.L. Novel Adociaquinone Derivatives from the Indonesian Sponge *Xestospongia* sp. *Mar. Drugs* **2015**, *13*, 2617–2628.
109. Guzmán, E.A.; Maers, K.; Roberts, J.; Kemami-Wangun, H.V.; Harmody, D.; Wright, A.E. The marine natural product microsclerodermin A is a novel inhibitor of the nuclear factor kappa B and induces apoptosis in pancreatic cancer cells. *Investig. New Drugs* **2015**, *33*, 86–94. [CrossRef] [PubMed]
110. Shih, S.P.; Lee, M.G.; El-Shazly, M.; Juan, Y.S.; Wen, Z.H.; Du, Y.C.; Su, J.H.; Sung, P.J.; Chen, Y.C.; Yang, J.C. Tackling the Cytotoxic Effect of a Marine Polycyclic Quinone-Type Metabolite: Halenaquinone Induces Molt 4 Cells Apoptosis via Oxidative Stress Combined with the Inhibition of HDAC and Topoisomerase Activities. *Mar. Drugs* **2015**, *5*, 3132–3153. [CrossRef] [PubMed]
111. Esposito, G.; Bourguet-Kondracki, M.-L.; Mai, L.H.; Longeon, A.; Teta, R.; Meijer, L.; Soest, R.V.; Mangoni, A.; Costantino, V. Chloromethylhalicyclamine B, a Marine-Derived Protein Kinase CK1δ/ε Inhibitor. *J. Nat. Prod.* **2016**, *79*, 2953–2960. [CrossRef] [PubMed]
112. Wang, R.; Zhang, Q.; Peng, X.; Zhou, C.; Zhong, Y.; Chen, X.; Qiu, Y.; Jin, M.; Gong, M.; Kong, D. Stellettin B Induces G1 Arrest, Apoptosis and Autophagy in Human Non-small Cell Lung Cancer A549 Cells via Blocking PI3K/Akt/mTOR Pathway. *Sci. Rep.* **2016**, *6*, 27071. [CrossRef] [PubMed]

113. Roel, M.; Rubiolo, J.A.; Guerravarela, J.; Silva, S.B.L.; Thomas, O.P.; Cabezassainz, P.; Sánchez, L.; López, R.; Botana, L.M. Marine guanidine alkaloids crambescidins inhibit tumor growth and activate intrinsic apoptotic signaling inducing tumor regression in a colorectal carcinoma zebrafish xenograft model. *Oncotarget* **2016**, *7*, 83071. [CrossRef] [PubMed]
114. Loaã, C.N.; Attanasio, E.; Villiers, B.; Durieu, E.; Tahtouh, T.; Cam, M.; Davis, R.A.; Alencar, A.; Roué, M.; Bourguet-Kondracki, M.L. Marine-Derived 2-Aminoimidazolone Alkaloids. Leucettamine B-Related Polyandrocarpamines Inhibit Mammalian and Protozoan DYRK & CLK Kinases. *Mar. Drugs* **2017**, *15*, 316.
115. Li-Chun, L.; Tzu-Ting, K.; Hsin-Yi, C.; Wen-Shan, L.; Shih-Min, H.; Tsui-Chin, H. Manzamine A Exerts Anticancer Activity against Human Colorectal Cancer Cells. *Mar. Drugs* **2018**, *16*, 252.
116. Simithy, J.; Fuanta, N.R.; Alturki, M.; Hobrath, J.V.; Calderón, A.I. Slow-Binding Inhibition of Mycobacterium tuberculosis Shikimate Kinase by Manzamine Alkaloids. *Biochemistry* **2018**, *57*, 4923–4933. [CrossRef] [PubMed]
117. Kumar, N.S.; Dullaghan, E.M.; B Brett, F.; Huansheng, G.; Reiner, N.E.; Jon Paul Selvam, J.; Thorson, L.M.; Sara, C.; Nicholas, V.; Richardson, A.R. Discovery and optimization of a new class of pyruvate kinase inhibitors as potential therapeutics for the treatment of methicillin-resistant Staphylococcus aureus infections. *Bioorg. Med. Chem.* **2014**, *22*, 1708–1725. [CrossRef] [PubMed]
118. Kuroda, I.; Musman, M.; Ohtani, I.I.; Ichiba, T.; Tanaka, J. Pachastrissamine, a cytotoxic anhydrophytosphingosine from a marine sponge, *Pachastrissa* sp. *J. Nat. Prod.* **2002**, *65*, 1505–1506. [CrossRef] [PubMed]
119. Kwon, Y.; Song, J.; Bae, H.; Kim, W.J.; Lee, J.Y.; Han, G.H.; Lee, S.K.; Kim, S. Synthesis and Biological Evaluation of Carbocyclic Analogues of Pachastrissamine. *Mar. Drugs* **2015**, *13*, 824–837. [CrossRef] [PubMed]
120. Frederic, M.; Williams, D.E.; Patrick, B.O.; Irwin, H.; Robert, M.; Kim, S.C.; Roll, D.M.; Larry, F.; Rob, V.S.; Andersen, R.J. Liphagal, a Selective inhibitor of PI3 kinase alpha isolated from the sponge *Aka coralliphaga*: Structure elucidation and biomimetic synthesis. *Cheminform* **2006**, *37*, 321–324.
121. Markwell-Heys, A.W.; Kuan, K.K.W.; George, J.H. Total Synthesis and Structure Revision of (-)-Siphonodictyal B and Its Biomimetic Conversion into (+)-Liphagal. *Org. Lett.* **2015**, *17*, 4228–4231. [CrossRef] [PubMed]
122. Zong, Y.; Wang, W.; Xu, T. Total Synthesis of Bioactive Marine Meroterpenoids: The Cases of Liphagal and Frondosin B. *Mar. Drugs* **2018**, *16*, 115. [CrossRef]
123. Zhang, J.; Ren, L.; Xi, Y.; White, M.; Greenhaw, J.; Harris, T.; Wu, Q.; Bryant, M.; Papoian, T.; Mattes, W. Cytoto xicity of 34 FDA approved small—Molecule kinase inhibitor s in primary rat and human hepatocytes. *Toxicol. Lett.* **2018**, *291*, 138–148. [CrossRef] [PubMed]
124. Medical Data Website of China. Available online: https://data.pharmacodia.com (accessed on 5 May 2019).
125. Muñoz-Alonso, M.; Enrique, Á.; María José, G.N.; Marina, P.; Pablo, A.; Galmarini, C.M.; Alberto, M.O. c-Jun N-terminal kinase phosphorylation is a biomarker of plitidepsin activity. *Mar. Drugs* **2013**, *11*, 1677–1692. [CrossRef] [PubMed]
126. Galmarini, C.M.; Maurizio, D.I.; Paola, A. Trabectedin and plitidepsin: Drugs from the sea that strike the tumor microenvironment. *Mar. Drugs* **2014**, *12*, 719–733. [CrossRef] [PubMed]
127. National Library of Medicine, US. Database of Clinical Studies. Available online: http://www.clinicaltrial.gov (accessed on 15 May 2019).
128. He, H.; Tran, P.; Gu, H.; Tedesco, V.; Zhang, J.; Lin, W.; Gatlik, E.; Klein, K.; Heimbach, T. Midostaurin, a Novel Protein Kinase Inhibitor for the Treatment of Acute Myelogenous Leukemia: Insights from Human Absorption, Metabolism, and Excretion Studies of a BDDCS II Drug. *Drug Metab. Dispos.* **2017**, *45*, 540–555. [CrossRef] [PubMed]
129. Stone, R.M.; Manley, P.W.; Larson, R.A.; Capdeville, R. Midostaurin: Its odyssey from discovery to approval for treating acute myeloid leukemia and advanced systemic mastocytosis. *Blood Adv.* **2018**, *2*, 444–453. [CrossRef] [PubMed]
130. Knapper, S.; Russell, N.; Gilkes, A.; Hills, R.K.; Gale, R.E.; Cavenagh, J.D.; Jones, G.; Kjeldsen, L.; Grunwald, M.R.; Thomas, I. A randomised assessment of adding the kinase inhibitor lestaurtinib to 1st-line chemotherapy for FLT3-mutated AML. *Blood* **2016**, *129*, 1143–1154. [CrossRef]
131. Bourhill, T.; Narendran, A.; Johnston, R.N. Enzastaurin: A lesson in drug development. *Crit. Rev. Oncol. Hematol.* **2017**, *112*, 72–79. [CrossRef]

132. Bernardo, M.L.; Belén, V.; José Esteban, P.R.; Arturo, S.M.; Juan José, P.R. Population pharmacokinetics of kahalalide F in advanced cancer patients. *Cancer Chemother. Pharmacol.* **2015**, *76*, 365–374.
133. Lefranc, F.; Koutsaviti, A.; Ioannou, E.; Kornienko, A.; Newman, D. Algae metabolites: From in vitro growth inhibitory effects to promising anticancer activity. *Nat. Prod. Rep.* **2019**, *36*, 810–841. [CrossRef]
134. Lien, W.C.; Chen, T.Y.; Sheu, S.Y.; Lin, T.C.; Kang, F.C.; Yu, C.H.; Kuan, T.S.; Huang, B.M.; Wang, C.Y. 7-hydroxy-staurosporine, UCN-01, induces DNA damage response and autophagy in human osteosarcoma U2-OS cells. *J. Cell. Biochem.* **2018**, *119*, 4729–4741. [CrossRef]
135. Pujari, R.; Jose, J.; Bhavnani, V.; Kumar, N.; Shastry, P.; Pal, J.K. Tamoxifen-induced cytotoxicity in breast cancer cells is mediated by glucose-regulated protein 78 (GRP78) via AKT (Thr308) regulation. *Int. J. Biochem. Cell Biol.* **2016**, *77*, 57–67. [CrossRef] [PubMed]
136. Andel, L.V.; Rosing, H.; Schellens, J.; Beijnen, J. Review of Chromatographic Bioanalytical Assays for the Quantitative Determination of Marine-Derived Drugs for Cancer Treatment. *Mar. Drugs* **2018**, *16*, 246. [CrossRef] [PubMed]
137. Barr, P.M.; Lazarus, H.M.; Cooper, B.W.; Schluchter, M.D.; Panneerselvam, A.; Jacobberger, J.W.; Hsu, J.W.; Janakiraman, N.; Simic, A.; Dowlati, A. Phase II study of bryostatin 1 and vincristine for aggressive non-Hodgkin lymphoma relapsing after an autologous stem cell transplant. *Am. J. Hematol.* **2009**, *84*, 484–487. [CrossRef] [PubMed]
138. Farlow, M.R.; Burns, J.M.; Gorelick, K.J.; Crockford, D.R.; Grenier, E.; Wilke, S.; Cooper, E.C.; Alkon, D.L. Bryostatin-1 improves cognition and daily living tasks in moderate to severe Alzheimer's disease: Preliminary report of a phase 2 study. *J. Alzheimer's Assoc.* **2017**, *13*, 1476. [CrossRef]
139. Keck, G.E.; Poudel, Y.B.; Rudra, A.; Stephens, J.C.; Kedei, N.; Lewin, N.E.; Peach, M.L.; Blumberg, P.M. Molecular modeling, total synthesis, and biological evaluations of C9-deoxy bryostatin 1. *Angew. Chem. Int. Ed.* **2010**, *49*, 4580–4584. [CrossRef] [PubMed]
140. Naman, C.B.; Rattan, R.; Nikoulina, S.E.; Lee, J.; Miller, B.W.; Moss, N.A.; Armstrong, L.; Boudreau, P.D.; Debonsi, H.M.; Valeriote, F.A. Integrating Molecular Networking and Biological Assays To Target the Isolation of a Cytotoxic Cyclic Octapeptide, Samoamide A, from an American Samoan Marine Cyanobacterium. *J. Nat. Prod.* **2017**, *80*, 625–633. [CrossRef]

MDPI
St. Alban-Anlage 66
4052 Basel
Switzerland
Tel. +41 61 683 77 34
Fax +41 61 302 89 18
www.mdpi.com

Marine Drugs Editorial Office
E-mail: marinedrugs@mdpi.com
www.mdpi.com/journal/marinedrugs

www.ingramcontent.com/pod-product-compliance
Lightning Source LLC
LaVergne TN
LVHW070637170726
843515LV00004B/272

* 9 7 8 3 0 3 9 4 3 7 8 3 2 *